# Solving Management Problems
# in Information Services

## CHANDOS
### INFORMATION PROFESSIONAL SERIES

Series Editor: Ruth Rikowski
(email: Rikowskigr@aol.com)

Chandos' new series of books are aimed at the busy information professional. They have been specially commissioned to provide the reader with an authoritative view of current thinking. They are designed to provide easy-to-read and (most importantly) practical coverage of topics that are of interest to librarians and other information professionals. If you would like a full listing of current and forthcoming titles, please visit our web site **www.chandospublishing.com** or contact Hannah Grace-Williams on email info@chandospublishing.com or telephone number +44 (0) 1865 884447.

**New authors:** we are always pleased to receive ideas for new titles; if you would like to write a book for Chandos, please contact Dr Glyn Jones on email gjones@chandospublishing.com or telephone number +44 (0) 1865 884447.

**Bulk orders:** some organisations buy a number of copies of our books. If you are interested in doing this, we would be pleased to discuss a discount. Please contact Hannah Grace-Williams on email info@chandospublishing.com or telephone number +44 (0) 1865 884447.

# Solving Management Problems in Information Services

CHRISTINE URQUHART

Chandos Publishing
*Oxford · England*

Chandos Publishing (Oxford) Limited
Chandos House
5 & 6 Steadys Lane
Stanton Harcourt
Oxford OX29 5RL
UK
Tel: +44 (0) 1865 884447 Fax: +44 (0) 1865 884448
Email: info@chandospublishing.com
**www.chandospublishing.com**

First published in Great Britain in 2006

ISBN:
1 84334 136 0 (paperback)
1 84334 184 0 (hardback)

© C. Urquhart, 2006

British Library Cataloguing-in-Publication Data.
A catalogue record for this book is available from the British Library.

The Publishers make no representation, express or implied, with regard to the accuracy of the information contained in this publication and cannot accept any legal responsibility or liability for any errors or omissions.

The material contained in this publication constitutes general guidelines only and does not represent to be advice on any particular matter. No reader or purchaser should act on the basis of material contained in this publication without first taking professional advice appropriate to their particular circumstances.

Typeset by Domex e-Data Pvt. Ltd.
Printed in the UK and USA.

# Contents

# List of figures

# List of tables

# Acknowledgements

Christine Urquhart thanks her family, colleagues and friends for their support during the writing of this book. Special thanks also go to the Chandos editorial team, particularly Dr Glyn Jones, and Neill Johnstone, for all their help and encouragement.

# Preface

One of the main reasons for writing this book was to explain some of the more interesting and more useful methods that library and information service managers could use to evaluate their services and plan for the future. In my role as an academic, I read the research literature and I realised that the literature was becoming more difficult for the practising library manager to assimilate, for several reasons. First, much of the literature that could be useful is buried within the management literature, and it takes some imagination and time to translate this into methods useful for library and information services. Second, and probably more important, the growth of what might be termed library and information science has been sporadic, but also highly specialised. The language used in many of the articles demands an excellent background in mathematics and statistics to understand the meaning fully, and few library and information service managers, in my experience, have that background.

I am not sure when the gaps between the specialist 'library scientist' or 'information scientist' and the generalist library manager began to emerge, but it seems that over the past 20 years some, though only some, of the researchers and the research work became divorced from many of the concerns of library practice. The qualitative research may have flourished at the expense of some of the quantitative research. The concern for 'relevance' obscures the problem of developing good research questions that can be answered from the mess of many practical problems. For development of the discipline, particularly in the UK, this divergence seems surprising as well as unfortunate. Much of the pioneering work on library problems was done by the founder of the National Lending Library for Science and Technology, Donald J. Urquhart. It was Donald Urquhart who became aware of the work done by statisticians on the skewed frequency distributions that are so common among library and information data collections. Using these distributions to analyse the data sets, Urquhart set about applying these to various library problems on collection management, purchasing and relegation. Bensman (2005) notes that Donald Urquhart was an advocate of numeracy among librarians, and in particular, not just the ability to do arithmetic but also an understanding of statistics, and how statistics can be applied to practical problems. Most of the UK library and information science lecturers have not, perhaps, succeeded in conveying such knowledge to their students. For this book I have tried to find examples in the literature for further reading, but it has been very difficult to find relevant examples that are accessible to those who are shy of more advanced mathematics, for whatever reason.

I think I should make it clear at this point that I am no relation of the famous Donald J. Urquhart, despite sharing the same surname and a training in the natural sciences. I also share an interest in mathematics and statistics, although it is many years since I did A-level mathematics. I have found it depressing over the years that librarians are, generally, very unwilling to engage with anything to do with numbers, as I strongly believe that many of the problems, as well as much of the research, demand a reasonable level of numeracy. That's not to deny the importance of leadership, and good human resource management, but to say that someone is a leader implies that there are people willing to follow them. I, for one, would like to be reassured that my leader could keep the finances in order, both now and in the future. Being warm, cuddly and caring is all very well, but services need to be cost-effective, and planned carefully, with due regard for the risks involved. Wouldn't it be better to take calculated risks and provide a service that thrives, rather than making do, and simply providing a service that limps along?

My main aim for this book is to provide an accessible introduction for some of the techniques that might be used by library and information service managers to review their services and plan for the future. I would also like readers to find some of the techniques fun to use, but that is an aspiration rather than an aim. I have tried to find examples for further explanation and I have also referred to further reading in management science textbooks that will provide further details. It is a rather eclectic mix, and inevitably it reflects some of the techniques I have used over the years and found useful. I may have omitted some techniques, and if that is the case, my apologies. I wanted to avoid writing the type of book that presumes the reader to be unintelligent as well as ignorant. I personally dislike the type of book that tells me what to do, without explaining why or setting out the evidence for and against ways of doing things. This is a 'how to do' book, but one written more to give you ideas on things to try or to speculate about. Research within the Department of Information Studies at the University of Wales Aberystwyth has indicated that many of the longstanding problems with training needs in the sector are numeracy problems (Urquhart et al., 2005). There is a common thread linking the lack of expertise in bid writing, performance measurement and benchmarking, and statistics for critical appraisal, and that is numeracy. I have been bemused by the concern about information literacy in the profession – information is not just contained in words but also in numbers, and images. Before preaching the virtues of information literacy for lifelong learning, perhaps we should be more confident in our own skills in interpreting numerical information, whether presented in statistical analysis or in simple spreadsheets.

The structure of the book reflects the eclectic mix of topics. Chapter 1 discusses performance indicators and performance measurement. The chapter is not intended to be a summary of previous books on performance measurement. The intention is to discuss what makes a good performance indicator, and how some of the data quality considerations affect the validity of performance measures. Chapter 2 examines some applications of game theory to the setting of standards and norms in library and information service dealings. Libraries have traditionally cooperated, but the basis of that cooperation should be questioned occasionally, and some of the problems that arise in purchasing consortia for electronic information can be viewed through the perspective of the familiar 'prisoner's dilemma'. There is also no harm done in re-examining the games of bluff and negotiation that may be played between those providing electronic information services and those purchasing electronic information services. There are strategies that serve both sides well. Chapter 3, also concerned with electronic information, discusses the attempts made to model the amazing growth of Google. These follow the tradition of earlier studies on the growth of the literature, and the models put forward to explain, and possibly predict future growth. Chapter 4 examines some of the methods used to evaluate the impact of information provided by a service on the user or users of that service. One of the concerns I have always had about impact evaluation is that managers should not think the job is done when the impact is measured and the results show, hopefully, that their service has some positive impact on the users' lives or work. That is just the start, as the impact evaluation needs to be fed into future service planning. The chapter includes a 'rough guide' to costing. Chapter 5 explains how information can be used in decision making, and future options assessed for their risk, costs and benefits. Is it better to play safe, or take the calculated risk? This chapter will not provide all the answers but it explains some techniques that can be used to think about the decisions to be made. Chapter 6 carries on the theme of future planning with some techniques for analysing the past, and predicting what might happen in the future.

# About the author

**Christine Urquhart** joined the Department of Information Studies at the University of Wales Aberystwyth in 1993, before which she worked as a library manager in a college of nursing. She has worked as an information professional in the pharmaceutical industry, the electricity supply industry and has also been a self-employed technical abstractor. She is currently Director of Research and she leads the research training postgraduate programmes. Her research interests in impact and performance measurement started in the college of nursing, and her first post for the University was as a researcher on an impact project in the health sector. She has always enjoyed research work as it involves solving problems, experience which has helped in her information systems and management teaching.

The author may be contacted at the following address:

Dr Christine Urquhart
Department of Information Studies
University of Wales Aberystwyth
SY23 3AS

Tel: 01970 622188
E-mail: *cju@aber.ac.uk*

# Performance measurement

Performance measurement is an important element in library and information service management. The difficulty for many information professionals is that the estimation of library performance is inextricably tied up with precious professional values. Challenge the professional values that are held very dear and the immediate reaction is often very defensive. Unfortunately, however valuable the service to the cultural heritage, funding tends to come only to those services that can demonstrate success in meeting funders' objectives, as cost-effectively as possible.

Library services can, and do measure an amazing number of inputs and outputs, such as the length of shelving and the number of items loaned, but measurement of inputs and outputs by themselves are meaningless. It might be tempting to be impressed by an extraordinary increase in activity, say double the number of sessions booked for literacy training. However, if you knew that the number of staff to teach those sessions had quadrupled, and the number of computers had increased by a factor of ten, then the increase in activity seems much less impressive.

This chapter focuses on performance measurement and performance indicators, particularly the measures, and how reliable these are. Chapter 4 discusses impact evaluation, and the questions around finding the value of a service, including some of the basic costing issues. However good the performance of the service may appear to be, someone may ask whether the service can be improved. At the end of the chapter there is a brief introduction to one variety of business process modelling.

## Definitions for performance indicators

There are some definitions that are important to get right:

- *Inputs*: What goes into the service, what is bought, donated, or acquired, and the human resources, the number of staff and type of staff.

- *Outputs*: What is produced by the service, such as loans, answers to enquiries, teaching sessions and document deliveries handled.

- *Outcomes*: What the service users do with the service outputs. The information service produces the outputs, such as training sessions for 'silver surfers', but it is the silver surfers themselves who are responsible for the outcomes. They, not the information service, will be using the Internet to make contact with friends and family, or to download and view family photographs. Outputs are relatively easy to assess, as it is easy to measure the number of training sessions or number of search sessions, but outcomes require good knowledge of customer behaviour. For assessing the impact of a library and information service, we need to know more about outcomes than outputs.

In performance measurement, the following should be remembered:

- *Comparison*. You should be comparing inputs with relevant outputs (books in, books out, reference staff available, enquiries handled) or inputs with outcomes (number of teaching sessions offered, student learning outcomes).

- There are *three traditional 'E's*: efficiency, effectiveness and economy:

  - Efficiency compares inputs with outputs – how much was produced in relation to the inputs?

  - Effectiveness compares inputs with outcomes – what, or how much did the customers achieve in relation to the inputs provided by the information service?

  - Economy compares the outputs with the inputs in financial terms. This is useful when trying to justify the investment costs of equipment.

- The fourth E is *efficacy*, the power to bring about a desired result. We can think about the way our services contribute to getting something done within our community or workplace. Just as important, we can think about the way the organisation of our services, including our support services, contributes to greater skills, knowledge or learning among our staff.

# Developing performance indicators

Performance indicators use some of the principles applied in cost-benefit analysis. It is often very difficult to put a figure on the value of a spectacular view – how do we measure beauty when what is beautiful to one person is not beautiful to another? We can get some idea of the value most people place on the view by comparing, for example, the price of a house with no view, with the price of a similar house with a splendid view. Similarly, we do not know what attributes of a web page appeal to particular individuals, but we can often assume, other things being equal, that the longer the time spent reading a web page, the more valuable it may be. The time spent reading is a proxy value for the 'benefits gained' in reading the web page.

We can measure inputs, we can measure outputs, and we can try to relate an input to an output and hope that it tells us something meaningful about the service performance. If service A provides more outputs than service B for the same inputs, then service A is more efficient than service B. Is it realistic to expect more outputs if more inputs are provided? That depends on the type of input and the likely timelag in ensuring that everything is up and running. The new member of staff (an additional input) may need time to learn about the job, before working to the same level of performance as those who have been in the post for some time.

We can also study trends in performance quite easily if we can assume that the changes in inputs have been minimal. In such situations, if outputs are increasing from year to year, then this indicates that performance is increasing.

The word *performance* reminds us of a stage performance, where to put on a performance requires not just the actors and the props, but support behind the scenes as well – the musicians, the lights, the costume manager, the box office staff. Similarly, when assessing the performance of a library and information service, it is important to remember that relating one input to one output may omit some quite important parts of the picture. To produce one particular output may require coordination of several inputs, and that output may relate to other outputs. We need to be careful in our choice of performance indicators, those relations between inputs and outputs. We need to be assured that putting this input with that output is a fair test of the performance, but we also need to check whether ups and downs in that performance indicator is telling us something sensible about the service. For example, in an impact study of nursing library services several years ago (Davies et al., 1997), we found that several library services were very proud of the number of interlibrary loans they provided for their nursing staff users. One of the reasons the nurses requested so many interlibrary loans was, however, the lack of nursing journals in those libraries. Was an increase in the number of interlibrary loans, in relation to the number of staff servicing that service a useful indicator of the performance of the service? How was it possible to compare the performance of this service with that of a library which had developed a collection of nursing journals, and which, therefore, did not field as many requests for interlibrary loans?

The performance indicator should be measuring something valued by users, namely access to resources. We can partially measure this through an indicator for interlibrary loans, or an indicator for the percentage of core nursing journals on subscription, but we need to remember that our measures are only hitting part of the target. The measurements may be very precise, but they may only be measuring part of what we need to measure. Both measures need to be considered in relation to each other. To develop useful targets we audited several libraries to be able to estimate, on the basis of several aspects of performance, what the realistic targets should be, and what the minimum (and maximum) numbers to be expected should be.

For example, on that project the targets that we developed were:

- for a library with more than 100 nursing journal titles: target average number of interlibrary loan request per requester = more than 2, but not more than 4;

- for a library with fewer than 50 nursing journal titles: target average number of interlibrary loan requests per requester = at least 3.

In an age of electronic journals these targets are obviously no longer meaningful, but there is a general point to be made. Those libraries with fewer physical resources should be measuring their efficiency in providing access to resources elsewhere. Those libraries with physical resources, priding themselves on their collection, should be extending access to other resources for their users, but their collection efficiency should mean that their target will be framed differently. Two library services may have the same numerical target, but for different reasons.

## Data quality for performance measures

One of the reasons why libraries have a vast mass of performance data is that some measures are relatively easy to collect. The number of visitors can be calculated automatically at a turnstile entrance, the number of items loaned, the number of reservations, the number of visitors to a web page – all these are simple to collect. The questions that need to be levelled at any 'performance measure', any item of data that is going to contribute to performance assessment of the service, are:

- How reliable is this data? Is the data collected one day as reliable as the data collected another day?

- How robust is the data? How easy is it to cheat, accidentally or on purpose?

■ Can we be sure that our definitions, of 'visitor', for example, relate clearly and unambiguously to the data collected?

Discussions with staff who are dealing with users or usage data will often reveal the problems in performance measures. Computerised data collection relies on equipment that does not break down, or develop faults. In the past, working in a small library, we had a library management system that counted a renewal as a new loan, but our definition of a loan did not count a renewal as a new 'loan'. This meant that we kept a manual tally of loans – not easy when the library was busy, and always liable to some mistakes when retrospectively doing the tally after a busy period. To make life simpler for ourselves, we might have collected the data generated by the system over a set period (a month, perhaps), and compared that with the manual tally. Six months of data collection might have allowed us to compare the manual tally of loans with the system tally of loans. If there was a reasonable correlation every month, then we might have applied a correction factor to the system tally and assumed that the adjusted figure would correspond, more or less, to the manual and presumably more accurate loan tally.

The correlation between the system tally and the manual tally can be calculated using a function (CORREL) on Microsoft Excel. The correlation comes out to 0.66, which is a reasonable indication that the data in each set are moving in line with each other. A large number for the system tally is matched with a large number in the manual tally and vice versa, see Table 1.1 and Figure 1.1).

If, however, there is one period in the year when there are very few renewals, the system tally and the manual tally will be much closer in that period, but not at others. Compare

| **Table 1.1** | Comparison between system and manual tallies |
|---|---|

|  | Jan | Feb | Mar | Apr | May | Jun | Jul | Aug | Sep | Oct | Nov | Dec |
|---|---|---|---|---|---|---|---|---|---|---|---|---|
| System tally | 3567 | 3400 | 3673 | 3496 | 3783 | 3275 | 3120 | 2976 | 3672 | 3794 | 3649 | 3418 |
| Manual tally | 3000 | 2900 | 3050 | 3175 | 3262 | 2978 | 2877 | 2541 | 2804 | 2953 | 3089 | 2930 |

**Figure 1.1**  Correlation one between system tally and manual tally

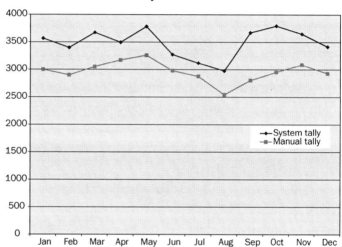

Figure 1.2 with Figure 1.1 and note how the spring and early summer period differ. Unsurprisingly, the correlation has decreased to 0.56. The value of drawing the chart is that it is easier to see that the pattern differs in one season of the year. If you were going to use the system tally with a correction factor, and the monthly variation was important to you, then you might be better to use two different correction factors, if the seasonal variation was a regular pattern.

When using Microsoft Excel to calculate some statistical functions it is worth checking what the calculation is doing

**Figure 1.2** Correlation two between system tally and manual tally

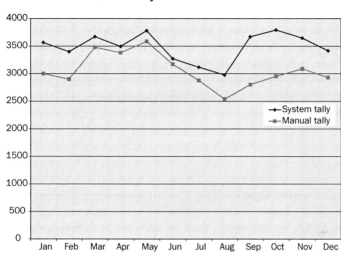

and what the assumptions are. The correlation function uses the covariance, and the formula includes the mean (average) loan figure for the year for both the system and the manual tallies. Is this reasonable? The average loan figure is something that is meaningful, and there will be variations throughout the year with some periods busier than others. We don't need to know why, but the correlation function calculation at least provides us with an indication whether the system tally is varying consistently with the manual tally. The maximum value the correlation function could have is one, which would mean perfect correlation, and the minimum value is near zero, which would mean there was no correlation between the two data sets.

The example of actual loans versus the system generated loan figures is one aspect of performance indicators – considering whether what you are measuring truly indicates the aspect of performance you want to assess, even if it is not possible or practical to measure it directly. The degree of similarity reflects how valid your chosen performance measure is. In research terms there are several types of validity, and some debate about how these should be defined. Strictly speaking, the type of similarity discussed in the loans example is a criterion type of validity, as it is more about prediction than explanation. We don't need to know exactly what the system is doing to calculate its loans figure, all we really want to know is whether we can be reasonably assured that we can apply a correction factor to the system loans figure to obtain a figure that is as close as possible to the real loans figure.

Another type of validity that is important in performance measurement is construct validity. Considerable research (Cook and Thompson, 2000a,b, 2001) has been done on instruments such as SERVQUAL, to find out how close some of the measures are to assessing user satisfaction with library services. There are two ways of thinking about this. First, we often want to know whether there are some factors that we could assess, and whether these tend to cluster together as a 'construct' – a collection of factors that we might think characterise 'lack of confidence' in a library user, for example. Another way of thinking about the clustering is that we might find that we have been happily measuring something in our routine library surveys for years. However, the coming year, we are told that we cannot ask our users that question for ethical reasons. For our own service purposes, we need to have some idea of the answer to that question, as it helps to predict the need for specialist services. Close inspection of our library survey results for the past five years suggests that there is another question that we ask our users that might help us. The answers to that question, and the answers to the question we can no longer ask, are strongly correlated. We therefore do not need to worry about the question we can no longer ask, as the question we can ask will be a good proxy indicator for us.

Tools such as SERVQUAL measure the importance of various factors on a scale of 1–5, from not important to very important. However, for the five dimensions of tangibles, reliability, responsiveness, assurance and empathy, you are immediately dealing with the fuzzier concepts of 'responsiveness' and 'empathy'. We need to be assured that the questions used in the SERVQUAL instrument really do assess reliability as most users would think of 'reliability'. Another difficulty with instruments such as SERVQUAL is no real fault of the instrument itself, but more a problem of the increasing expectations of our users. If many of the factors are 'important' or 'very important' then we are not really using the full range of the instrument and we end up considering whether the difference between a rating of 4.5 and 4.3 is worth worrying about. A study in health libraries (Martin, 2003) shows that most service quality dimensions were important, and the only dimensions that were rated below 3.9 in importance were '24-hour access to resources' and 'visually appealing materials associated with services and resources'.

The validity of test measures is something that has been researched in the educational field, which is the place to look for further discussion about the definitions and types of validity that should be considered. If the thought of using a proxy indicator instead of 'real' indicator to measure something of interest worries you, then you should be reassured if you read how you can check whether or not you can do this. Often the problem is not in the statistics but in human nature and the problems of chasing targets. Once general aims are known and the proxy indicators assessed, together with the targets for these proxy indicators, the effort to meet the targets can distort the entire activity of a service. We might meet the targets, but it is likely that the targets are no longer properly assessing what they were meant to be assessing in the first place.

## Comparing performance indicators

We can measure inputs, we can measure outputs, and we can try to relate an input to an output and hope that it tells us something meaningful about the service performance. This is one way of developing performance indicators for comparing two different services. For example, imagine Library A provides 2,000 loans per full-time library assistant and Library B provides 4,000 loans per full-time library assistant. Clearly Library B seems to be more efficient than Library A, as far as loans are concerned. This is only part of the performance assessment of the service, as all parts of the service need to be considered, but it is an indicator nevertheless. Comparing performance indicators across different services depends, naturally, on good quality data. The data used for the performance measures for inputs and outputs need to be reliable and robust, and the libraries need to ensure that they agree on the data definitions, which might need to be quite detailed. What happens about staff who are on long-term sick leave, and who have not been replaced? Library staff are unlikely to be enthused by performance indicators that indicate they have been slacking when the real reason that the productivity appears low in comparison with other services is that the calculations are based on staff on the payroll, with no account taken of sick leave.

For individual services, the year to year trends are also of interest. We can also study trends in performance quite easily if we can assume that the changes in inputs have been minimal. In such situations, if outputs are increasing from year to year, then this indicates that performance is increasing. However, if there are changes in inputs, such as an increase in staff numbers, then an increase in output might be expected. So, beware of simple comparisons of performance measures such as number of loans year on year.

# Measuring quality through sampling

It may be easy to measure the number of visitors through the security gate in the library, and although you may believe the machine has a few temperamental days, this measurement is something you do not generally need to worry about. But what about questions concerning the quality of the cataloguing? It would be impossible to go through every single item, but how do you devise a fair sample, and what sort of things should you measure?

Ann Chapman and Owen Massey (2002) describe a catalogue audit tool that was developed for the University of Bath. They looked at the twin sides of the problem, going from the catalogue to the collection, and from the collection to the catalogue. This should be done by printing a random sample of catalogue items, finding the corresponding items on the shelf (catalogue to collection) and then checking the item five places to the left of that item (collection to catalogue). The researchers found it was impossible to generate a true random sample, and the first convenience sample study examined the catalogue records for several hundred books issued over one weekend. Obviously, as the researchers note, items in multiple copies can cause problems to the procedures used, and a rush on a group of items for an assignment could bias the sample. If an item is inaccurately catalogued and shelved where it should not be, then the likelihood of it being borrowed by someone browsing along the shelves in a chosen subject area is decreased. It is important in this type of setting to look at major errors and minor errors. As the researchers found, it is important to be clear about the meaning of data quality in this type of exercise. Incomplete information is not necessarily the same as missing information. For their second systematic sample, the researchers used systematic sampling of the shelves. This was justified, they argued, as the first convenience sample test had indicated that the items selected on the shelves were all on the catalogue.

## Performance indicators for electronic information services

One of the difficulties with assessing the effectiveness of digital library services is finding suitable measures. In the early stages of web page development, the number of hits was often taken as a proxy indication of the popularity of the website, but that quickly became discredited, once it was clear that people might simply be clicking on to the page by mistake. If people really valued the information on the web page, they might save the information or print out the page. They might spend longer on the page, although that is assuming they are working on the page – they may have been interrupted or they may have gone off to make a cup of coffee. In one research project I spent hours looking at the searches being conducted by remote users, and tried hard to figure out what was happening. They rarely sat down and did a search from start to finish as I had innocently expected. In fact, as I was switching between several concurrent users to see what was going on, the monitoring period used to be characterised by long periods when no user was doing anything very much, and short periods of frantic activity by a couple of users, which was difficult to record accurately, as this was well before the days of extensive transaction log analysis.

Transaction log analysis has indicated that information behaviour is not the same as it might be in the print environment. What does it mean to see the evidence of the 'bouncers' and 'flickers' (Nicholas et al., 2004) – is this good or bad? It depends, partly on the intentions of the website sponsors, but, to some extent, our metrics (Nicholas et al., 2001) need to understand or try to understand the behaviour first, before assuming that a long time reading is 'a good thing' and that quick browsing and checking is not.

## Process indicators

If performance indicators suggest that a service is not as efficient as it might be, is it possible to do further assessment of the process? There are several levels at which this question should be considered. At the top level, the question is about the business processes and their interrelationships. Do we really need to look at the entire business of the organisation, to look at the processes, the roles involved and the business rules? Processes are not necessarily to be equated with the service departments or sections. At the next level, we are more concerned with the operational management, who does what, where the interrelationships and responsibilities lie, how things happen and in what order. At the next level of detail we are looking the detail of procedures within a process.

Business process modelling is more than the time and motion studies that were associated with aspects of operations research. There are several approaches to business process modelling and it is worth looking at several guides to find one that suits your service and suits you. One approach that does link some of the strategic and operational aspects, in a very detailed and rigorous way is the business process management RIVA approach developed by Martyn Ould (2005). There are several key definitions in this approach:

- process is a coherent set of actions carried out by a collaborating set of roles to achieve a goal;
- a role is a responsibility within a process;
- an actor carries out a role;
- a role carries out actions according to business rules;
- a rule has props which it uses to carry out its responsibility;
- roles have interactions in order to collaborate;
- a process has goals and outcomes.

In this approach, there are two different relationships between processes: activation, starting off another process, and interaction, which may be a request or delivery type of interaction. The approach recognises that there should be three different type of processes:

- the case process (typically, what the staff are doing);
- the case management process (typically, the supervisory aspects);
- the case strategy process (typically, the process that should be there to maintain a strategic view).

Much of the performance measurement literature is concerned with the operational or supervisory level. Impact and its implications for strategic planning might be considered strategic.

To take one example, let's see how this might work for a process management problem. Webb and Galloway (1999) describe process management and performance measurement of a collection development problem in a university library. The library was perceived to be not very efficient in supplying known items from reading lists. The research involved interviews with the staff involved in the collection management activities, as well as analysis of the time involved in the various stages of getting books available. The various stages identified were order processing, order approval, supplier-related activities, database team activities, and subject team checking. Variations in process times were measured, and this showed that although the mean process times were reasonably satisfactory, there was a wide variation between the high and the low times for some of the teams. The researchers identified two problems they viewed as fundamental – and different. These were bottlenecks and variations in process time due to bad process design.

Using the basic principles of the RIVA process (for further details consult Ould, 2005, Chapters 4 and 5) we need to think about 'units of work' involved in the collection development activities. The article indicates 'item receipt' and 'checking items received' as likely units of work.

If we define a case process as 'handle receipt of items' we can view the outcomes of this case process as, possibly, being to approve receipt, or to reject, or to query the receipt. There is also an associated process, to 'handle the checking of items received'. The framework that is used in the RIVA architecture is presented in Figure 1.3, where an order (A) generates items received (B), and in this case, an item received (A) generates receipt checks (B).

**Figure 1.3**   **Units of work and their relationships**

A Unit of Work 'A' generates Units of Work B

An order (A) generates items received (B), and in this case, an item received (A) generates receipt checks (B).

For a general service relationship between two units of work, Figure 1.4 (slightly simplified from the original RIVA instructions) shows how to set about looking at processes and their relationship. Figure 1.4 illustrates the framework, and Figure 1.5 shows what the diagram would look like when illustrating the relationship between receiving items (item receipt) and checking items received.

Figure 1.4 shows a general service relationship that could exist between two units of work. The case process for the unit of work requested by A, interacts with the case management process for unit of work B. The case management process is 'managing the flow' of the case process. In the service relationship there is a link between

**Figure 1.4**   **Service relationships between units of work**

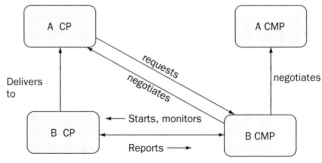

CMP: case management process; CP: case process

**Figure 1.5**    Handling items received, and checking receipt

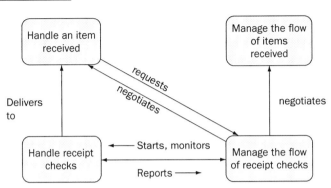

the case management processes for A and B, but this can be simplified to a 'task force relationship' which means that the top right-hand process box for 'A CMP' disappears.

For the example, the process illustrated in Figure 1.5 can be described as follows. Handling an item received means that we need to perform some checks on the item, to ensure that the item is in good order and what was ordered. Managing the process of receipt checks requires someone (or a staff role) to initiate the checks, and there may be actions such as monitoring, intervening and stopping the process, as well as reporting back to the monitoring process on progress or other problems. The outcome is that the checking will deliver something to the process of receiving items. If some of the work for dealing with items received has been outsourced, then it is likely that we need to illustrate a service relationship. This is because someone needs to be responsible for managing the flow of items received, which will require some interaction with the management of the flow of the checking. If the entire operation for these two processes is performed inhouse, then we can essentially encapsulate the management of the flow of the receipt checking within the process of receiving items.

The figures simply illustrate the case process and the case management process for the unit of work. Each unit of work should also have a 'case strategy process' which is essentially concerned with observing long-term performance and trends, making predictions and determining future strategy for the case process.

There is not enough information in the article by Webb and Galloway to show how the problems in process design, or the bottlenecks might be identified. To illustrate the process in detail and the interactions requires another set of role-based diagrams. Using these, and the information in the unit of work figures summarising the process relationships, analysis might provide answers to questions such as:

- What is the critical path (the sequence of activities that must be done to ensure completion of the process)? Are there activities on this critical path that are only of internal value to the organisation? If so, can they be moved off the critical path to ensure that the critical path focuses on value to the customer?

- Are there activities done in sequence that could be done in parallel?

- Do we have too many 'specialist' roles when a generalist role might save time overall, through greater flexibility? Can we identity any roles that are simply paper shuffling between two other roles?

As you have probably gathered by this stage, full business process modelling may require some highly structured thinking about the processes occurring in any library and information service. Simply measuring the performance indicators for efficiency can lull a manager into thinking that all is well, but if asked how to improve the performance, some clear thinking about processes may be necessary. If we are concerned about the efficacy of our service, thinking about the processes and how they might be improved offers one way of empowering the frontline staff to analyse what they are doing and how the operations could be improved.

# 2

# Cooperation and collaboration

One of the problems encountered by many designers of knowledge management systems is that the users simply do not want to cooperate by sharing expertise with others. How does the situation appear to the users of the knowledge management system? Game theory allows us to model the payoffs to cooperating and non-cooperating users. This is commonly known as the prisoner's dilemma. This chapter explains some of the ways in which game theory might be applied to library and information service management problems. Social network analysis is another tool that has been used to study the exchange of information between people.

## The prisoner's dilemma

The basic outline story of the prisoner's dilemma is that two prisoners have agreed, before capture by the police, that neither will say anything to the police about the offence they may be charged with. Neither has committed an offence, but the police have enough evidence for conviction for a minor offence, but not enough for a more serious crime. The prisoners are questioned separately. If both prisoners stay silent, both will only be convicted for the minor offence. The dilemma occurs when the police offer each prisoner a reduced prison term if they confess to the serious offence and give evidence against the other prisoner. For an individual prisoner, this offers them a tempting prospect. If they defect on their friend, their payoff is good, but for their friend the prospects are not at all good. If both defect, the outcome is less good than mutual cooperation (as they are both accusing each other of the more serious crime). If you and a friend were prisoners, what would you do? If you defect, and give evidence against your friend, the payoff for you looks very tempting. If you both keep your word, then the individual outcome is less good. If you both defect, then you will both be punished. The dilemma is whether you can trust your friend to cooperate with you and not to defect.

The usual layout of the matrix for illustrating the payoffs is shown in Figure 2.1.

By adding up the total payoffs, you can see the combined payoff if both you and your friend cooperate (3 + 3 = 6), and if you both defect, the combined payoff is much less (1 + 1 = 2).

In real life, typically in many knowledge management systems, people have repeated opportunities to cooperate

**Figure 2.1**  Prisoner's dilemma matrix

Your payoffs

Other player

|  |  | Cooperate | Defect |
|---|---|---|---|
| You | Cooperate | 3 | 0 |
|  | Defect | 5 | 1 |

The other player's payoffs

Other player

|  |  | Cooperate | Defect |
|---|---|---|---|
| You | Cooperate | 3 | 5 |
|  | Defect | 0 | 1 |

with each other (or not). In a very ideal world, repeated cooperation would lead to a very rewarding outcome for all, but the point of game theory is that you are operating as an individual and trying to maximise the reward for yourself.

Imagine playing ten rounds of this game. If you and the other player consistently cooperated, the combined payoff would be 60, and each of you would gain 30 points individually. But the maximum you could gain individually would be 50 points, assuming you defected all the time and the other player cooperated every time. This, however, is unlikely, unless they are the complete sucker every time. Try playing one of the games that are available on the Web to see how you might get on in repeated rounds of the game.[1]

Although the total reward for mutual cooperation is greater than the sum of the other payoff totals, for the individual there is a dilemma. It could be in their individual interest not to cooperate, particularly if the next player chooses not to cooperate. If they chose to cooperate, and next player does not cooperate, then the first player becomes the sucker. If they strongly suspect the other player will not cooperate, the safe, but boring option is not to cooperate at all. In such cases, there is a punishment for mutual non-cooperation, but it is not substantial. The lesson for knowledge managers is that the rewards of mutual cooperation need to be highly visible, and large, or that the punishment for mutual non-cooperation is large, perhaps by making such cooperation almost obligatory.

## Evolving strategies and norms behaviour

In many situations we need to consider how the prisoner's dilemma will evolve over many rounds of play. Is there a

strategy that works well for the individual? Studies (Axelrod, 1997: 33) suggest that a generous tit-for-tat strategy is effective, even when it is difficult to know whether the other player is making a deliberate choice or has just made a mistake. In other words, do what the previous player has just done. If they cooperate, then you cooperate, and if they defect (do not cooperate) then most of the time you should not cooperate either, but you should be prepared at times to be generous and cooperate. Similarly, if the other player defects after you have defected, then you need to be contrite, and change your strategy to cooperate some of the time.

There is a variation in the prisoner's dilemma game that considers what happens when there is a known chance of being observed. If the player does defect (or cheat), then there is a chance of being detected, and if the other players see the defection they may choose to punish the defector. If the defector is punished then the cost of punishment is considerable, but the punisher also has to pay a cost, as it is costly to them to enforce the punishment. As a practical example, for the defecting player read student who plagiarises, and for the other player read tutor who has to pay a cost in time in detecting the plagiarism and enforcing the punishment.

If the student does not cheat, then there are no extra rewards, and no enforcement costs on the tutor. If the student does cheat, then they get a payoff of +3 points (using the same scheme as illustrated in Axelrod, 1997: 49). As they have done better, in relation to other students as a result of the cheating, the other students or players in this game are hurt, not by much perhaps, but their reward is –1 point. Now there is a chance that the tutor will observe the cheating, and if they do, the student who is observed to have cheated may be punished severely. If that happens, the payoff for the cheating student is –9 points. Unfortunately, it takes time to detect that cheating, and the cost to the tutor of enforcing the rules is –2 points. There are variations in this game, and because of the chances of being observed cheating, the payoffs will vary. The vengefulness of the tutor may vary, and they may choose not to punish the student. More generally, the game can be played on the basis of other students (players) being able to observe the cheating sometimes, but even so, one can assume that there will be an enforcement cost.

Let's imagine one round of this game, with ten players. One player (John) cheats (defects) and gets a payoff of +3 points. Each of the other players are hurt, and they each get a payoff of –1 point. Mary observes the cheating by John and decides to punish him, so he gets –9 points, but Mary has to pay the enforcement cost of –2 points. Another player, Paul also observes the cheating but decides not to punish John. The total scores at the end of round one are:

- John: –6 points
- Mary: –3 points
- The other eight players each have –1 point.

John has paid a penalty, but so has Mary, for being vengeful for the sake of the common good. The example illustrates just one round of the game but a simulation can be done over several rounds, to determine how strategies will evolve over time, and how different values for vengefulness (willingness to punish) and boldness (willingness to cheat) affect the long-term outcomes. A little imagination suggests that in the next round Mary is unlikely to be looking out for possible cheats, and in fact, the other players are unlikely to cheat as John paid a penalty and they know Mary observed the cheating. However, you can imagine that in a later round of the game, one of the bolder players might reckon that their chances of being observed are quite low, and even if they are observed, they may not be punished. The potential rewards are tempting for them individually.

Robert Axelrod (1997) discusses various ways in which game theory can be applied to the processes underpinning the development of standards and norms, the incentives for firms to join one alliance or another, and the way cultural regions develop and change in size over time. John Maynard Smith (1982) examined how game theory could be applied to an evolutionarily stable strategy for animals whenever the best thing for them to do depends on what others are doing.

An interesting example from animal behaviour that has some relationship to negotiations in the business world is the situation when there is incomplete information about the resource to be shared. How much is a library service prepared to pay for a bundle offered by a publisher? The price increase demanded is higher than the library really wants to pay, but the publisher is prepared to accept a mid-way position (although the library does not know this), rather than face the possibility that the library cancels its subscription completely. Game theory calculations can be applied to the situations of bluffing (library threatening to cancel the subscription, in the hope that the publisher reduces the price) and negotiation (offers from both sides in the hope of finding a happy compromise position).

There are other theories that can be applied to purchasing decisions of electronic information services and products (Urquhart, 2002). Most of these include some consideration of the trust and power relationships involved. One area of concern to many organisations, trade unions and national governments is that of outsourcing. Various conceptual frameworks have been applied to outsourcing, and it is important to remember that there are various types of outsourcing as well. For purchasing of electronic information, pricing bands may be used by purchasing consortia. Game theory reminds us to consider the perspective and the potential payoff for each member of the consortium. Although librarians may prefer to cooperate with each other, they also have to ensure that the interests of their institution are served as well as possible. Small institutions in any consortia may be cooperating to maximise their access to resources, as a way of surviving as best they can. Larger institutions may be more concerned with competing with other institutions (as research-intensive institutions generally do) as cooperation is fine for them if it offers a cost-effective way of plugging gaps in their resources. The goals are different for the two different types of institution. Although it is in everyone's long-term interests to cooperate, to present a united front, there is a temptation to defect.

## Social network analysis

The principles of social network analysis[2] concern:

- relationships (content, direction and strength);
- ties (between actors).

Analysing a social network is very much a 'bottom up' approach to identifying whether or not we have groups, rather than assuming that because teams and groups are created according to some organisational diagram, teams and groups will function as such. The 'grouping' emerges from the analysis of the direction of the relationships, the pattern of the direction and the strength. The strength of a tie can be assessed in different ways, but the strong ties indicate actors who are highly connected, with frequent communications and information exchange between them.

The analysis of social networks mathematically depends on graph theory, to deal with the direction of the relationships and the patterns that emerge.

There are several ways in which sociograms can be represented, and many textbooks on organisational behaviour will illustrate examples of 'communi-grams' or equivalent. For example, at a meeting, the following types of communication could be represented:

- question;

- statement;

- reflection (answering own question – oh I have just realised that ...);

- attempt to speak, not noticed by the chair.

The direction of the question can be illustrated. Most should be directed to the chair of the meeting but there may also be some communication between members of the committee at the meeting. There are frameworks for group interaction that could be used to break down this categorisation further but this would require tapes or video recordings to help consider whether a statement was a proposal, a supporting statement or a disagreement.

The density of a network gives some indication of the extent of information exchange. Haythornthwaite (1996) discusses some examples that show different degrees of centralisation, and density (Figure 2.2).

The highly connected network A is also quite democratic as there is equal communication between all members of the network. Network B is more like many bureaucratic, divisionalised structures with a managing clique around a centre (director) and communication between the members at the fringes going through an intermediary. Network C could have been drawn as a straight line as communication

**Figure 2.2**   Network examples

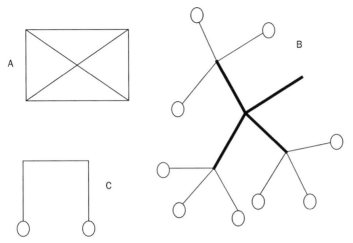

*Source:* Haythornthwaite (1996)

from one end to the other has to go through the intermediate points. There are no other connections, unlike network A.

# Notes

1. See, for example, *http://www.economicsnetwork.ac.uk/archive/ poulter/pd.htm* (last accessed: 17 January 2006) and J. E. Beasley's 'OR-Notes'; available at: *http://people.brunel.ac.uk/ ~mastjjb/jeb/or/contents.html* (last accessed 18 January 2006).
2. See, for example, the NHS National Knowledge Service (2005) 'Social network analysis'; available at: *http://www.nelh.nhs.uk/ knowledge_management/km2/social_network.asp* (last accessed 17 January 2006). This provides details of the software tools available.

# 3

# The Pareto rule and the problems of user satisfaction

In this chapter we will examine why Google became so popular so quickly, and some of the network models that illustrate how this might happen. For marketing of library and information services there are some lessons that can be learned about viral marketing. The Pareto 80:20 rule usually means that 80 per cent of the use of a library service comes from 20 per cent of the potential users in the user population. We look at the background to this rule, and whether it is possible, or even desirable to find a way around this when trying to reach customers that do not appear to want our services.

## The growth of Google

When scientists such as Albert-László Barabási started mapping the Web to examine how it worked as a network, his team considered various network theories. For more details about the models and the approaches used by Barabási, consult the book *Linked: The New Science of Networks* (Barabási, 2002).

The random network model, for a large network, predicts that almost all nodes in the network will have approximately the same number of links. In other words, if the network was the human population, the random network model predicts that most people will have the same number of links to other people, and that we will all have approximately the same number of friends or acquaintances. A few people will be highly social or atypically antisocial. For websites, a random network model predicts that most sites will have the same number of visits (see some examples of networks in Chapter 2). Is this idea of most sites having the same number of visits true? Experience tells us that while many sites are not visited frequently, some are visited more than others.

Other network models attempt to model the small-world feeling commonly known as 'six degrees of separation', that it is possible to link any two randomly chosen people through five links, of five other people. So A knows B, B knows C, C knows D, D knows E, and E knows F. For a community such as librarians, it is probably not necessary to have as many as five links. Think of the latest new member of staff that arrived in your library or information service, and remember the introductory conversations you had with that member of staff. I'd take a bet that you could find a chain of at most five people linking you with that new member of staff, even if you had never met them before. It is likely that the chain is even shorter, with one mutual acquaintance in common. Barabási describes many discoveries of the six degrees of separation by different researchers. For example, Milgram's study asked randomly chosen residents of small towns in the middle of the USA to send on a letter to someone they knew that might be more likely to know the target American identified by Milgram. The recipients were asked to inform Milgram's research group when they had received the letter, and who they were, so that Milgram could keep track of the chain. And, lo and behold, for the letters that made the complete journey, the median number of intermediate persons was 5.5.

Around the same time, another sociologist, Mark Granovetter, was looking at social networks from a different perspective. What interested Granovetter was the importance of our acquaintances, in a variety of social activities. Typically we have a close network of strong ties with friends who are friends with each other, however, we, and our friends, will also be connected through 'weak ties' with other friendship groups. Finding out about that important vacancy, or good flat that is about to become vacant, is often achieved through our friends' acquaintances, rather than our immediate friends. Two mathematicians, Watts and Strogatz, illustrated this as the long-range links between distant nodes (Figure 3.1).

Both the random network model and the small-world cluster model that Watts and Strogatz developed, seem to explain some features of society as we know it, but the type of clusters they predict differ. The random network model emphasises that most of us have approximately the same

**Figure 3.1**   Small world network

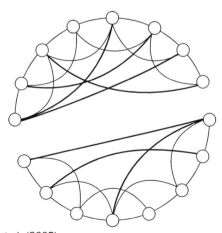

*Source:* Huang et al. (2005)

number of friends and acquaintances, but that some people may be more highly connected. The small-world cluster model emphasises the importance of a few long-range links that connect distant nodes, offering short-cuts that make the 'six degrees of separation' possible.

Barabási's team was interested in the structure of the Web, but the map of the Web produced by their robot crawler for their university's domain did not seem to match either of these models well. Around 82 per cent of the pages on the Notre Dame domain have three or fewer incoming links, but a small minority had been referenced by many other pages, and had over a thousand incoming links. Looking wider to the Web itself, it was obvious that there were hubs, connectors that linked to a large number of sites. Instead of dismissing the connectors as atypical, the research now focused on the behaviour of these hubs, asking why web pages and many other types of networks had such hubs. Look at an airline map when you are next on aeroplane, and compare the type of map shown there with the average road map linking towns on the ground. The airline map will probably show hubs, which are connected to many other airports. The small airports may be connected to only a few other airports.

The importance of the people nodes has been recognised in management studies for a long time, although described in different ways. When researching what made teams successful, Meredith Belbin (1993) identified one essential component of the team as the 'resource investigator', the type of person with lots of connections, a strong social network beyond the immediate team. These findings echo those of Allen , who identified the 'information gatekeeper' in his work on technology transfer among science and engineering teams (Allen, 1977).

Try this exercise with around 20 (at least) of your colleagues, asking them how many professional societies or committees they belong to (preferably voluntarily). You may find this a useful exercise in finding out how many connections you have externally, but the principle of the exercise is really in the shape of the graph you will generate.

Plot the number of societies along the horizontal x-axis (starting at zero and finishing at the maximum number of societies identified). Plot the number of people belonging to each number of societies up the y-axis (vertical axis). As you can see, on the example curve shown (Figure 3.2) 18 people belong to one society, with a steep decline after that, as only two people belong to three societies, two to four and one to five societies.

I cannot predict the shape of graph that you will generate, but I'd expect it to be more similar to the power law shape than the bell curve shape. The important thing to remember about the power law curve is the 'long tail'. Note also that with the bell curve we could easily calculate the average number of links, and say that most nodes will have a certain

**Figure 3.2** Society memberships

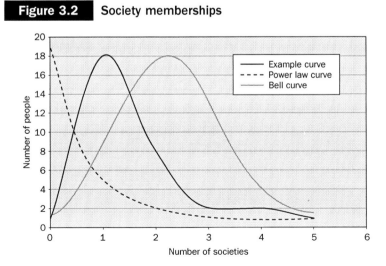

number of links. For the power law distribution there is no such thing as a characteristic node. This is shown in the figure, as there is no clustering around a middle number. Very probably, a few people will be very social and belong to lots of professional societies, and most people will belong to one or two at most. The very well connected people are the 'hubs', the type of people you know to approach when you want to know someone's address or something about someone, as these are the people who are likely to know.

How do the hubs, the nodes with large number of connections get to be hubs? How did Google become so pre-eminent in such a short period of time, in other words? When the research team at Aberystwyth on the JUSTEIS project examined the behaviour of staff and students throughout the UK on the use of electronic information services, Google was just mentioned by a few interviewees of the 200 interviewed in 1999, and most of the interviewees mentioning Google were librarians. By 2004, Google was a household name. Clearly, the word was beginning to get around in 1999 and the results were impressive. As one of our students interviewed said, 'This Google ... it's incredible', although there were other search engines available at the time and Yahoo had the advantages of being earlier in the field. It is much easier to accept that the first mover obtains certain advantages of being first established. Had AltaVista, Yahoo or one of the other early search engines been so successful that they had crowded out all the other competition, there would have been no chance for Google to overtake. However, as our interviewees in JUSTEIS indicated, between 1999 and 2004, Google just got fitter than most of the other search engines. Our interviewees would use other search engines, often Yahoo, but over the years, it was clear that nearly all the students would use Google, and possibly another search engine. It seemed that when asked for advice, most people would say, 'try Google'. They might mention another search engine, but Google was also very likely to be mentioned. It didn't cost the recipient anything to try, it was easy to use, and successful results almost guaranteed. Similarly, on JUSTEIS we found in 1999 that our students were very difficult to contact by e-mail. We were using their university e-mail address but we later discovered that most of the students were using Hotmail accounts which they could access at home as well as at university. Hotmail was easy to use, free, and, of course, an e-mail sent from a Hotmail address advertises itself to other potential users.

What are the lessons of network models for the planning and development of library and information services? It is inconceivable that novice users would find a database service as easy to use as Google. The advanced intricacies of Boolean searches do not, on the whole, seem worth the effort. But getting a precise search result is useful, if not vital, for specialist researchers, and it is these groups that need advice and support, appropriate to their level of need. We need to keep an eye out for alternatives to Google, as the power law distribution with the long tail, reminds us that other hubs may emerge, and may operate as well, if not better, some day. Google Scholar and SCIRUS (from Elsevier) may emerge as new hubs, and may attract a dedicated following of users.

# Fostering networks and communities of practice

Returning to the social networks of Granovetter, (see also Chapter 2) the importance of fostering weak links is obvious. How can library services help here? One advantage of virtual communities of practice is that they make learning from the experience of similar, though not geographically close groups, much easier. You may even have experienced working in an organisation where two departments were working on similar projects and encountering similar problems, with no idea that other people might have had a solution to offer. Having a structure which encourages the sharing of knowledge is part of the mission of most knowledge managers. This may be a structure which encourages physical meetings of members from different teams and departments, or a virtual structure, posting case histories on a virtual community space to allow users to compare experience. It does not mean that such links are automatically made, but it makes the long-range links easier to maintain.

Alfredo Moreno (2001) describes how the communities of practice emerged at the International Development Bank, helping to make links between geographically distant professionals who were working on similar projects. These communities of practice were supported by coordinators who tried to bring in members and acted as coaches, to help the learning of the network members.

Information specialists working for the National Library for Health's specialist libraries look after virtual communities of practice that support specialist clinical interests, such as child health, musculoskeletal conditions and women's health. Indeed, the information specialists have their own informal community of practice, to share and exchange knowledge among themselves on how to organise content and foster the virtual community activities, as an evaluation of the project indicated (Cooper et al., 2005). Much of the information specialists' work is to ensure that the coordination of activities happens, that the collaborators for the various specialist libraries contribute their expertise, and that expertise is shared appropriately. They are ensuring that the links are in place for the network, ensuring that the 'weak links' are strengthened.

Ravid and Rafaeli (2004) studied asynchronous discussion groups to assess whether they conformed to one or more of the network models. Their definition of the small-world network examined the degree of clustering and the distance needed to pass from one node to another. In a small world, everyone is connected and a large group, like a village, is formed. In a scale-free network, like the Web, there is no characteristic node, but hubs, and a large number of nodes with only a few links. From the perspective of those wishing to measure intensity of usage of the discussion group, it is of interest whether links are being used rather than simply made once, and then available but not used further. For the examination of network models, only the presence or absence of a link is assessed, not the intensity of usage. Ravid and Rafaeli found that the discussion groups in a university's managed learning environment did behave as a scale-free network, although there were more smaller hubs than are found in other scale-free networks. Interestingly, the ten most active people in the network were not all tutors, as one might expect. Only two of the ten most active were tutors, the other eight were students who had many contacts. The network also behaved as a small-world network with a high degree of clustering. This study has implications for our expectations of the behaviour of students within managed and virtual learning environments. We tend to be concerned about the lurkers on the system, those who do not appear to be very active. If the hub network is the stable social structure, then perhaps we should not be worried if this is what appears. There is some order in the structure, and the degree of clustering obtained in a small-world network shows that, just as in a village, there may be people who do not appear to mix as much as other people, but the gossip reaches them just the same. We can start asking questions about the number of smaller hubs we would expect to find, centred on particular tutors, and we can analyse the behaviour of the connections to distinguish between networks that have a very few highly connected hubs, and those with more hubs and semi-hubs. But it seems that we should not expect everyone to have the same number of links to other people.

Another way of identifying other groups, but more often individuals with similar interests to your own, is obvious when you use the Amazon website. I am always amused at the problems Amazon has with our family's varied interests. This really stretches the collaborative filtering mechanism to its limits, trying to cope with a profile of interests that stretches from military history through systems analysis to Sophie Kinsella and music interests that are far too diverse to profile sensibly. However, what Amazon does successfully, for most individuals, is to link disparate individuals with similar interests and foster niche interests in a way that has been very difficult up till now. Libraries have done this by starting to group books on science fiction together, although many have just lumped all the science fiction together, not realising that the science fiction enthusiasts have very select tastes, and preferences for particular authors.

## J-curves and the Pareto Law

The reverse J-curve occurs in other studies of user behaviour. Wolfram (2003: 128) examined several studies of the number of terms used in search engine requests, and the most popular number of terms submitted (the modal number) was two terms per query. Very few enquirers submitted more than five terms per query, and so the familiar picture of a very sharp downward slope to the J is obvious, with a long tail with a few enquiries showing more terms. Strictly speaking, this curve (Figure 3.3) may have a small kink at the top. Mathematicians have tried various equations to try to find one general type of formula that will produce the different curves, for different values of the variables in the formula.

There is a general pattern with many of the 'information production process' distributions, as Wolfram has called these, and that is the reverse J-shape with a long tail. One of the difficulties in curve fitting is deciding what is more important – finding an equation that is meaningful or an equation that neatly fits the observed data? It is possible to devise a formula to get the curve to fit, but explaining what the variable might mean practically can be difficult.

The economist Vilfredo Pareto was interested in various economic inequalities, such as the distribution of wealth in a community. He noticed that about 80 per cent of the

**Figure 3.3**   Reverse J-curve

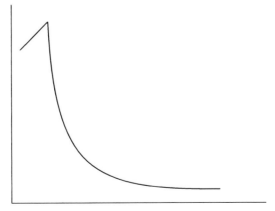

wealth was held by 20 per cent of the population, and found that the 80:20 rule applied to many other situations. Wolfram (2003: 49) cites some other examples of library book transactions, and that managers of academic libraries would probably find that around 80 per cent of the interlibrary loan requests come from 20 per cent of the academic staff. In the audit phase of the value project, I found that if I arranged the requesters by number of requests made, the top 10 per cent (those making the most requests) made, on average, 45 per cent of all the interlibrary loan requests. In general, it was quite obvious that a small active minority made heavy use of the service, and this was a common pattern across all the hospital library services, large and small. I puzzled about the implication of this at the time, but common sense suggests that library services make themselves very vulnerable to the vagaries of the few if their service becomes very heavily weighted towards serving a few key customers. At the time, I was interested in assessing whether charging for interlibrary loans made a difference to the number of requests and the distribution of the requests. There was insufficient data to test the hypothesis, but comments made by interviewees suggested that in some libraries some users did not want to overburden the library service, and therefore made fewer requests than they really needed. In others, the heavy requesters took it as their given privilege to send in large numbers of requests. There was some evidence to suggest that making a nominal charge deterred the heavy users from abusing the service, and encouraged the shy to make more requests.

The Pareto law is formulated as:

The number of people with wealth $W$ is proportional to $1/W^E$

Pareto found that $E$ varied between 2 and 3, no matter which country he surveyed, and this type of relationship holds for other situations as well.

Another way of formulating the Pareto law is:

$$r = \frac{A}{y^a}$$

where $r$ is the number of members of a community who earn more than $y$ dollars, and $A$ and $a$ are parameters ($a$ is the same type of parameter as $E$ in the other equation).

If we change this to the number of users making more than ten reservations (for ease of calculation $a = 2$), we would have:

Number of users making more than ten reservations annually $= \dfrac{A}{100}$

Number of users making more than one reservation annually $= A$

Number of users making more than five reservations annually $= \dfrac{A}{25}$

You might like to try this with your own reservations or perhaps your interlibrary loan requests to see whether this type of pattern holds true. Remember that you may need to alter the value of a, to get a better fit to your own data.

# Bibliometric studies

Those managing library collections or making decisions about which journals to retain and which to cut, are usually concerned with the measurement of which items in the collection are used most, and which can be cut, as they are used least.

One of the earliest studies was by Bradford (1934, 1948), who found that there was a law of literary yield for the dispersion of articles on a particular topic. First, rank the periodicals in order, with the periodical producing the largest number of articles on a particular topic first, and the periodical with the second largest next. Second, divide the total number of articles found by three. Third, mark off along the ranked list of periodicals how many periodicals contribute the first third of articles, then the second third. Bradford found a general law that no matter what the topic was, there was a relationship between the number of periodicals in each set. If there were $a$ periodicals in the most productive third, the next most productive set would contribute $a$ times $b$, and the least productive third would contribute $a$ times $b^2$. The values of $a$ and $b$ would differ but the relationship would hold. For example, if there were 30 periodicals in the first third ($a = 30$), and 90 periodicals in the next most productive set ($b = 3$), the least productive third would comprise 270 periodicals ($30 \times 9$).

Bradford's law is one of the power laws that have been developed. Another law also noted by Fairthorne (1969) was developed by Lotka to explain author productivity. The number of authors who had published $n$ papers in a given field was approximately $1/n^2$ the number of authors who had published only one paper. As Rousseau (2005) notes, Fairthorne was one of the first to realise that there could be a universal principle underlying all these various laws, and later researchers in informetrics tried to find the relationship between the distributions (Egghe, 1985, 2005a). Wolfram (2003: 46) cites Price's law, which says that the number of prolific authors in a subject area (producing around half the publications) is approximately equal to the square root of the total number of authors in the field.

As you might imagine, some of these laws raise some questions when trying to apply them. For example:

- Would there be more low productivity journals for a topic that was essentially multidisciplinary?

- How could the Lotka law be adjusted fairly for multi-authored papers, and is such adjustment necessary?

- Do the different Lotka $n$ values for different disciplines really mean anything?

- Are the models dependent on the time period chosen, and what is the effect, for example, of the research assessment exercise on publication patterns of UK scientists?

- Can the models cope with changes in the growth rate of the literature?

- How can differences among disciplines' preferred formats for publication, for example, monographs or journal articles, be included?

It is not within the scope of this book to discuss many of these ideas further, but simply to point out that some researchers have tried to answer some of these questions, and to develop more exact distributions which can model more of the distributions found in practice. For example, Shi Shan (2005) introduces a generalised Zipf distribution which has been used to model the distribution of web pages according to the numbers of links they have received. The basic Zipf law states that if $f(k)$ is the number of authors with productivity $k$, then the logarithm of $f(k)$ will vary linearly with the logarithm of $k$. Taking logarithms on both sides of an equation is (sometimes) a way of converting a relationship that is a power law (and which appears as a curve) into a linear relationship.

More exact distributions often mean more complex and lengthy functions as more variables and additional parts need to be added to allow for all the variations found in the models (see Wolfram, 2003: 80). A few of the more relevant examples are discussed in the following sections, to illustrate some of the ways in which bibliometrics might help library and information service managers.

## Assessing the research productivity of a university from its website

One example that compares whether one type of distribution was better than another at fitting the observed distribution of web links is the study by Payne and Thelwall (2005). This study examined in more detail the relationship between the links a university website creates or attracts and its research productivity. Calculation of a Web impact factor has in the past assumed that the relationship between the links to a university and its research productivity should follow a linear trend – if university A has twice the number of links that university B has, then A is about twice as productive in research as B. However, the distribution of links to an individual page on the university site follows a power law, with a few pages with a high number of links, and a large number with very few links. Payne and Thelwall (2005) assessed whether a power law was a better fit than a linear trend, as they had assumed. The specific question addressed was whether, for UK university websites, the relationship between site size and research productivity, and the relationship between outlinks and research productivity, should be modelled by a linear trend or a non-linear power law. Dealing with real-life data is messy, and there are always outliers, in this case sites that don't sit neatly on a straight line however it is drawn. The research showed that both the linear trends and the power laws fitted the data sets well. The authors conclude that a linear relationship seems justified for most of the relationships tested, although for some of the relationships a power law might be a better fit.

Judit Bar-Ilan's team examined the possible reasons for links between academic websites, concentrating on the source pages and the types of relationship (Bar-Ilan, 2005). Using this framework, a pilot analysis of Israeli university web pages indicated that the most common type of source pages (in descending order) were compilations (lists), contributions (actual information), units (about a physical place or group, or about a course or project) and persons. The distribution of target pages was different, with the most common type being a person, followed by unit and contribution. Many links confirm the hub and spoke nature of websites, with links from lists pointing to sources that carry the actual information. The most common type of relationship was 'is A', followed by 'provides information', 'is related/is similar', 'is created by', 'provides services and 'is useful/is relevant'. More research is necessary to validate the framework and to assess whether such work can aid the quantitative link analysis in helping to focus on the links that matter.

## Electronic journals and obsolescence

Although the development of electronic journals means the physical weeding of the collection is not the same problem that it was historically, user preferences for seeking out older material gives some indication of the value assigned to electronic archives. It is important to remember that presenting users with the ten most recent articles in a search may bias their preferences for more recent material, and that removing the older journals from the shelves is again no indication that users may find the material less relevant. It may simply be a measure of convenience. If it is convenient it will be used, if it is not convenient, only the most dedicated researcher will pursue the material. User behaviour may simply reflect what libraries assume users need, not what users really need.

First, we need a definition of obsolescence. The usual measure is the 'half-life', the time in which one-half of the currently cited literature was published.

The definition used by the Institute for Scientific Information is:

> Cited half-life is a measurement used to estimate the impact of a journal. It is the number of years, going back from the current year, that account for 50 per cent of the total citations received by the cited journal in the current year. (*http://scientific.thomson.com/support/ patents/patinf/terms/*)

In general, articles in the physical sciences age more quickly than those in the social sciences, and those in the social sciences more quickly than those in the arts and humanities, but this is a broad generalisation and there are differences between the disciplines and sub-disciplines that cannot be explained easily using this rule. For example, there may be the seminal articles, older articles that everyone cites. With the development of electronic archives, the tendency to cite the older material may increase as such material is easily accessible. Analysis of the transaction logs of EmeraldInsight (mainly management and business journals) and Blackwell's Synergy, over specific time periods showed that the download half-life was one year (Nicholas et al., 2005). Of a dataset which included nine months of 2003, most of the article downloads (over 50 per cent) were for articles published in 2002 or 2003. Other aspects examined in the study were:

- distribution of PDF and HTML use by publication year;
- table of contents use and article downloads;
- disciplinary differences in use of table of contents and article downloads, and use of articles by year of publication and by subject.

The data indicated that there were differences between disciplines but also between journal titles within the same discipline – some titles age faster than others. Examination of the EmeraldInsight data showed that there are different types of user as well. The practitioners and professionals tend to read only the journal articles published within the last year, whereas researchers have a greater appetite for the older material. It is early days with this type of research, but the indications are that making the archive more easily accessible does not necessarily result in greater use overall – there may be an effect with some journals and some types of user, but it is difficult to detect general trends. There was little evidence that 'big deal' subscribers, those with the greatest opportunity to access journals they may not have been able to access before, were using back issues any more than other types of user. Making the material available does not mean that users are queuing to use the material, and for some users, some disciplines and some journals, other approaches will be necessary to ensure that some users do not lose out if there is a cut-off date of 10–15 years for the electronic archive.

## Impact factors

Egghe (2005b) has discussed the effect of using different versions of the impact factor formulae. The two-year impact factor of a journal, as defined by the Institute for Scientific Information is:

$$IF \equiv \frac{c(1) + c(2)}{P(1) + P(2)}$$

where at time $t = 0$ (usually this year, but other years are possible); $c(1)$ is number of citations given by source journals in year 0 to articles of this journal published one year ago; and where $p(1)$ is the number of articles published in this journal one year ago. Similarly, $c(2)$ is number of citations given by source journals in year 0 to articles of this journal published two years ago, and $p(2)$ is the number of articles published in this journal two years ago.

As Egghe points out, the two-year period for a journal ignores the different patterns of obsolescence between disciplines and among journals within one discipline. If the ISI impact factor was a valid measure of impact it would not vary much from one year to the next, and observations suggest that there are changes. Egghe suggests that weighted Lorenz curves of a continuous variable would get rid of some of these problems, and others (e.g. Burrell, 2005) have worked on similar ideas with Lorenz curves for modelling informetric data. Egghe and Rousseau (2006) also contest that Lorenz curves provide a formal link between information retrieval and information sciences. Lorenz curves provide a better way, perhaps, of defining the similarity (in relevance) of two documents.

# Uses of Lorenz curves

The basic idea behind Lorenz curves[1] is that cumulative percentage frequencies are plotted against the class intervals.

For example, imagine that you wished to examine the effect of a different charging regime for document delivery requests on your customer base. You might want to examine which groups of customers, segmented by the number of requests made annually, contribute most to the library costs, and whether charging a higher or lower amount to those who make fewer requests would make a difference.

Table 3.1 presents a highly fictitious set of figures for the number of requests and the contribution to costs. I have assumed that the average cost may vary within each group, as some requests may be more expensive to obtain. If we plot the Lorenz curve, the graph we obtain looks like Figure 3.4.

This shows that nearly 80 per cent of the requests contribute about 60 per cent of the costs of the document delivery service. About 20 per cent of the requests contribute about 40 per cent of the costs. A charging regime which limited customer demand for the more expensive requests might be worth considering. If the requests and costs were

**Table 3.1** Cumulative requests and contribution to costs

| Requests up to | Frequency | Total requests (%) | Cumulative requests (%) | Costs | Total costs (%) | Cumulative costs (%) |
|---|---|---|---|---|---|---|
| 10 | 62 | 42.8 | 42.8 | 3,100 | 19.9 | 19.9 |
| 20 | 35 | 24.1 | 66.9 | 2,625 | 16.9 | 36.8 |
| 30 | 15 | 10.3 | 77.2 | 2,520 | 16.2 | 52.9 |
| 40 | 25 | 17.2 | 94.5 | 5,000 | 32.1 | 85.0 |
| 50 | 5 | 3.4 | 97.9 | 1,250 | 8.0 | 93.1 |
| 60 | 3 | 2.1 | 100.0 | 1,080 | 6.9 | 100.0 |
| Total | 145 | | | 15,575 | | |

**Figure 3.4**  Lorenz curve

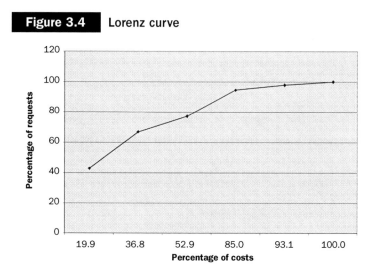

equally distributed, 10 per cent of the requests would account for 10 per cent of the costs, 50 per cent of the requests would account for 50 per cent of the costs. If you drew in a line of 'equal distribution' this would be a diagonal from the origin to the 100 per cent point at the top right. This curve is quite a long way from that diagonal, indicating that there is some inequality in the distribution.

## Final thoughts on user satisfaction

Most of the distributions that turn up in resource use, retrieval terms (e.g. indexing exhaustivity) have the reverse J-shape with the odd kink, reminding us that the Pareto rule covers many distributions of interest to library and information service managers. In attempting to show ever-greater demand for services, we may obscure (particularly from funders and policymakers) that the most active

20 per cent of users account for around 80 per cent of service use. Instead of going all out for growth, perhaps the emphasis in library and information service development should be on fostering the niche communities, the new hubs, whether these are groups of subject specialists in the workplace or the reading groups in a public library. We also need to find better ways of targeting our efforts more effectively, and the Lorenz curves indicate one approach for this.

## Note

1. See, for example, Wisniewski (1997: 58).

# 4

# Impact and cost evaluation

This chapter discusses some of the ways in which the impact of services can be measured. First, it is necessary to outline some of the basics of costing, as there is little point in assessing the impact of a service that cannot be justified economically in the first place. Chapter 1 explained some of the data quality issues that have to be considered in the performance measurement, and the introductory section covered the basic definitions. This chapter addresses the outcomes of information service use, the impact of information provided on the user's work, life, family and community, whether the community at work or where they live. Although the chapter emphasises the quantitative aspects of impact, such as time savings, this does not mean that some of the social and cultural impacts are excluded. The chapter begins with a short survival guide on costing.

## Basic costing

Many books deal with this in more depth, but to save time, the basics are reproduced here as an introduction to thinking about ways in which we cost savings and benefits for the users of library and information services.

First, we need to think about the inputs to the service, what we need to provide a service. We need space and offices, even for virtual library services. We need to buy resources, whether print or electronic. We need staff, who are a very important and valuable, but often expensive, part of the service. We need paper for the photocopier, ink cartridges, coffee for the coffee machine, and we need to maintain our equipment and buy replacements when the old equipment is out of date, or has broken down.

There are two main types of cost that arise when we are running the service, and two main ways of thinking about them. We can think of the costs as:

- First type of division:
  - *fixed*, loosely, we have to pay these whether our service is used a lot or used a little, or
  - *variable*, these vary according to the intensity of usage, for example, intensive usage of the photocopier means that we buy more paper and ink.

- Second type of division:
  - *indirect costs*, loosely, the type of costs that have to be paid for all types of functions that are offered, or
  - *direct costs*, costs that can be attributed directly to particular services, and which vary according to the amount of usage made of that service.

In practice, and for most cost estimates, it is probably easier to equate the fixed costs with the indirect costs, and the variable costs with the direct costs, although in accountancy terms there are differences.

## Rationing versus priority setting

One of the reasons for conducting an impact study should be to help in planning future services. There should be a virtuous circle of planning for service provision, implementation, auditing, and then making necessary changes (the plan-do-check-act methodology).[1]

The difficulty in assessing the impact when resources are not limitless, is that difficult decisions have to be made about doing one thing rather than another. We cannot manage demand for many library and information services simply by raising prices, as our services are provided free, or apparently so to the service users, at the point they access the services. Many users do pay, through their taxes, or there may be a charge to their department if the information service is in a commercial organisation, but there may not be a direct cost to the service user at the time they borrow or ask for assistance in an enquiry. If demand for any product exceeds supply, and supply is limited by resource constraints, then some type of rationing is inevitable, to ensure the customers receive 'fair shares'. We prefer to call this 'priority setting' but it is, effectively, rationing.

Klein (1996) discusses seven types of rationing, with reference to the health services, but equally applicable to many information services. These are rationing by:

- denial (excluding specific categories of client);
- selection (of those most likely to benefit);
- deterrence (making would-be users less welcome);
- deflection (making it someone else's problem);
- delay (the waiting list);
- dilution (spreading services thinly);
- termination (stopping services).

These strategies may be used singly, or in combination by those delivering the service. Often the strategies may not even be consciously applied, as they may have evolved, step by step, as a way of making life more tolerable for the staff when under pressure. Far easier to promise a little bit for everyone than to consult properly with user groups and reach amicable decisions? The difficulty of the 'spreading services thinly' option is that decisions to terminate services seem unreasonable to user groups who may, justifiably, wonder why they are being singled out, when everyone has been receiving an equally diluted service.

Research by the Department of Information Studies at Aberystwyth conducted among nursing libraries (Davies et al., 1997: 133–4) found plenty examples of the types of rationing that fitted the framework. Most of the time, these decisions were not conscious attempts at rationing, merely a response at one time to a particular problem. The long-term consequences were not important at the time. No type of rationing is better than another type but it is probably important to identify:

- type of rationing practised;
- how that rationing is put into practice, and how easy it is to operate;
- advantages and disadvantages, in particular the possible long-term damage.

The moral is that even if you do not think you are 'charging' for as service, there are hidden costs to the users that should be considered.

# Charging and modelling effects on demand

This section will provide a basic guide to some of the economic terms associated with studies of price and demand for services or products. It is basic and more details can be found in books such as Kingma (1996) and Butler and Kingma (1998). There are no easy routes around the economics of information. The market for information and information services is complicated by the ease of spreading information, and the fact that it may take years to produce a document, report or program, but it costs next to nothing to reproduce it. A structure needs to be set in place to compensate producers for their intellectual efforts otherwise there is no incentive for people to produce or provide these items. Hence we have copyright and patent legislation.

The demand for a product depends on the price set, and we know that in most circumstances the relationship between price and quantity sold will show that as the price rises, fewer items will be sold. If consumers have more income available to them, they can afford to buy more, and the quantity sold will increase as income levels increase. Just how much more, or how much less depends on the *elasticity* of the demand. If the highly esteemed professor on the library committee has to have his favourite journal, whatever the price, then the demand curve is not very elastic. This means that a price increase does not significantly lessen the demand. If, however, a price increase had a large and immediate effect, reducing demand, the demand curve would be elastic.

The concept of *externality* is another economic concept that may be useful when considering the effect of services on users. Kingma (1996, Chapter 5) discusses the example of junk mail. This may have a benefit (positive externality) on people other than the direct consumers of the junk mail, if they (unusually) like receiving mail. Mostly, we view junk mail as having negative externalities as we have to spend time scanning it to ensure that we are not missing something useful. The people who obtain the most direct benefit from junk mail are the firms advertising the products, who get more sales as a result of the advertising. Externalities are something to be considered when decisions about priority setting or rationing of services are made.

There are two main types of cost for cost-benefit calculations. The *marginal cost* is the cost of producing one extra unit of output, in terms of the variable costs involved. For example, producing one extra leaflet or one extra photocopy requires more paper and more ink. The costs of ink and paper depend on the number of leaflets or copies produced: they vary. There are also *fixed costs*, as the printer or copier has to be housed somewhere; there may be rental and maintenance costs as well as the overheads of heating and lighting for the room: these have to be paid whether the copier or printer does one copy or a thousand copies. The true annual *average cost* of producing one photocopy has to take both the fixed costs and variable costs into consideration.

You have probably already noted some of the problems in deciding whether a cost is variable or fixed, as rental and maintenance charges may vary according to the amount of work the machine does.

For a simple example, let's imagine we have the following annual fixed costs for operating and housing the copier:

- rental – £2,500;

- overheads – £2,000;

- staff – £18,500.

The variable costs are the paper and the ink. For ease of calculation let's assume that for every batch of 5,000 sheets of paper plus ink purchased, these are:

- paper – £50;

- ink – £50.

On the first day of the financial year you have to set aside the annual fixed costs, plus the payment for the first batch of paper and ink.

As Table 4.1 indicates, the first page copied costs a substantial amount, as you have to set aside the annual fixed costs plus the payment for the first batch of paper and ink. In fact the average cost, if you did no more than one copy, is £18,600. The marginal cost of the second page is the additional cost of the next page which is zero, as there is no extra cost in copying the second page. Thus, the average cost zooms down to a mere £9,300. This pattern continues, with the marginal cost of zero, until you are on sheet 5,000, at which point you have to load up with more paper and ink. The marginal cost of sheet 5,001 is £100 as you have to open up a new pack of paper and ink that costs £100. Note how the average cost increases slightly after the 5,000 mark, and again at 10,001, and 15,001.

| Table 4.1 | Copy costings | | |
|---|---|---|---|
| Number of copies | Marginal cost (£) | Total cost (£) | Average cost per copy (£) |
| 1 | 18,600 | 18,600 | 18,600.00 |
| 2 | 0 | 18,600 | 9,300.00 |
| 10 | 0 | 18,600 | 186.00 |
| 4,999 | 0 | 18,600 | 3.72 |
| 5,000 | 0 | 18,600 | 3.72 |
| 5,001 | 100 | 18,700 | 3.74 |
| 10,000 | 0 | 18,700 | 1.87 |
| 10,001 | 100 | 18,800 | 1.88 |
| 15,000 | 0 | 18,800 | 1.25 |
| 15,001 | 100 | 18,900 | 1.26 |
| 20,000 | 0 | 18,900 | 0.95 |

If you were asked to ensure that the price charged per copy was to cover costs, and that the price was to be 10 pence per copy, Table 4.1 indicates that at some point between 15,000 and 20,000 copies the average cost should become £1 per copy. Indeed, for copy number 18,900, with a total cost of 18,900 the cost is £1 per copy.

Although we could carry on with this type of table, to find the point where the cost becomes 10 pence per copy, it might be simpler to make an estimate that ignores the 'bumps' of the marginal cost at each 5,000 point.

We estimate that, for the paper and ink involved, each 'copy' costs:

variable costs per copy $= £100/5,000 = £0.02$

(2 pence per page)

price we need to charge $= £0.10$.

The contribution each copy makes to covering our fixed costs is not 10 pence, as each copy costs 2 pence for paper and ink costs. The contribution per copy is $10 - 2 = 8$ pence.

Our fixed costs are £18,500. The number of copies that we need to produce to break even can be calculated as:

(number of copies to break even) × (price per copy less the variable costs of producing a copy) = £18,500

$$\text{Number} = \frac{£18,500}{£0.10 - 0.02}$$

$$= \frac{18500}{0.08}$$

$$= 231,250 \text{ copies}$$

Therefore, we need to sell 231,250 copies to cover our costs (fixed and variable) if we price a copy at 10 pence per copy. Obviously if we charge more per copy, the breakeven point will be at a lower number of copies.

# Immediate impact assessment

Often we want to find out about the long-term impact, how the service users actually used the information they obtained and what difference it made. This is the holy grail: finding out whether the information in the interlibrary loan actually made a difference to the care of an individual patient, or whether the new electronic journal collection contributed to a successful grant collection, or whether the books provided made a difference to a older person's day, and helped stave off their feelings of loneliness. These are the important impacts we need to assess, but there are other immediate impacts that may seem less glamorous but are still useful.

The health sector impact studies conducted by the Department of Information Studies at Aberystwyth adopted and adapted a set of questions originally used in the Rochester impact studies (Marshall, 1992). There is a temptation to think that your set of impact questions need to be different, and original. That assumption is misguided, as it is far better to use a set of questions that have been tried and tested. Information professionals need to be more organised about providing sets of validated questionnaires that others can use, with minor modifications for their user groups and setting.

The questions (Urquhart and Hepworth, 1995: 247) are simple, but also very powerful. For example, with choice of answers of Yes, No, or Not applicable, 'what was the immediate impact of the information provided on your knowledge?'

- It refreshed my memory of details or facts.
- Some if it was new to me.
- It substantiated what I knew or suspected.

- I could use some information immediately.
- I will need to obtain more information on the topic.
- I expected to find something else.
- I will share this information with colleagues.
- I will add this to my personal information collection.

Answers to some of these questions are a good start to assessing the value of the information and the information service, in providing immediate answers (saving time for the user), sharing information with others (knowledge and information exchange, and secondary impacts of the information provided, reaching a wider customer base), and the value of decreased risk, by checking that the proposed decision or course of action was correct.

## Benefits in time saving

One of the more well-established methods of establishing the benefits or value of an information service is to estimate how much time the new service saves, compared with the traditional way in which information may have been obtained by the users.

In a preliminary evaluation of a networked database service, as part of the VIVOS project (Yeoman et al., 2001), a cost analysis was done to assess whether the service saved staff time. The assumptions made were that there were two groups of users. The first group were substituting time they may have spent searching in other ways for their use of the networked service. The second group were new users, who might not have previously have searched for information, but who were attracted to using the new service. This may not seem an important distinction to make, particularly when library services believe that their users should be making use of new services provided, but it makes a difference to calculating the *opportunity costs*, the purchase price plus the value of time spent consuming the item, in this case the time spent searching the database. The total costs associated with use of the database need to include the time spent retrieving the articles and then the time spent reading them. The results of the VIVOS study did show that there was an overall cost saving, but a closer look at the figures and the pattern of costs shows that most of the cost savings could be attributed to a few high earning individuals who used the service very frequently. For most of the users the service did not appear to provide many savings at all, and for those users who were learning how to use the service, the time spent could be more than they had previously spent on obtaining the information.

As indicated earlier in the chapter, most library and information services ration services to cope with demands greater than available resources. One of the ways in which demand for services can be managed is to impose a user fee, that may not cover the costs of providing that particular service (such as document delivery) but which acts to damp down demand to a reasonable level. Few studies illustrate how demand for services might vary according to the price set, although Mark Kinnucan (1993) examined whether price or speed of delivery was the crucial factor in determining preferences for one or the other. Unsurprisingly perhaps, price was the most important consideration then, but that may not be true now. Kinnucan used conjoint analysis, which compares consumer preferences for a product or service in terms of combinations of different attributes. The attributes assessed in the study were method of ordering the document, mode of delivery, speed of delivery and price. This type of analysis is useful in determining which attribute is most important, and whether, for example, the need for an urgent article might make people, or some user groups, more willing to pay a higher price.

Kingma (1998: 145) examines whether it is more economical to use interlibrary loans or provide access to electronic journals via subscriptions. The cost matrix includes, for the library users, the costs of filling out forms, waiting for articles to be delivered and any user fee. If the journal is available, then the costs involve finding the article and copying it (or printing it out if available full-text). Staff time costs, whether dealing with interlibrary loans, or dealing with journal delivery, shelving or 'monitoring' also need to be calculated. A similar analysis (Kingma, 1996: 156) illustrates how a change in journal subscription price makes it more economical to use interlibrary loans (or share resources in another way).

Library service managers might wonder whether open access journal articles are truly free for their users. Calculating cost per use depends on the definitions of use and costs. Holström (2004) has reviewed various studies of cost per article reading, noting some of the practical difficulties, such as estimating as 'one reading' in the situation when a reader views an HTML file and then downloads the PDF file. This should be counted as one reading if done at one sitting, but in fact looks like two downloads. Open access journals are, of course, not free, as they need to be supported by upfront fees levied on authors, or authors' sponsors. To calculate the cost per use of Biomed Central for the University of Helsinki, Holström simply divided the membership fee by the number of 'uses', the full-text downloads. The real cost per reading is more than the cost per use calculated on the number of downloads, and it is assumed that a figure of 2,000 'uses' or downloads really represents only 1,500 readings (i.e. 75 per cent of the uses approximately equates to user 'readings').

## Final thoughts on impact assessment

S. R. Ranganathan's Five Laws of Library Science, provide a useful summary of the main considerations of impact assessment and performance measurement. The Laws are:

1. Books are for use.

2. Every reader his book.

3. Every book his reader.

4. Save the time of the reader.

5. The library is a growing organism. (Ranganathan, 1963)

We can use these principles for networked services and electronic journal bundles just as easily as they might have been used for traditional printed collections. Collections should be relevant to the needs of the user population, there should be something available and accessible for an individual user, and niche uses should be considered. The user's time is often overlooked but needs to be carefully considered in our cost-benefit calculations. We should consider the externalities, and the hidden costs of rationing. Finally, we need to adapt our provision and methods of evaluation as services evolve.

## Note

1. See *http://www.isixsigma.com/me/pdca/* (Last accessed 18 January 2006).

# 5

# Information and uncertainty

Classically, the provision of information is meant to improve decision making by reducing uncertainty. By providing more information it is assumed that decision making will be easier. Library and information professionals tend to assume that it is a good thing to provide our users with information to help them with their studies, work or hobbies. Of course, we realise that providing them with quality information is better than providing rubbish, but we assume that 'information is a good thing' and that our users should certainly pay in time for using our services and the information provided, even if we do not directly charge them for the information. This chapter deals with decision making under uncertainty. This includes assessment of the likelihood of certain outcomes and how much we may regret our decisions should things not turn out the way we think they might.

## Informational decision analysis

How could we model the value of information in a simple setting? An example from Hirshleifer and Riley (1992: 185) illustrates how this could be calculated for a simple betting example.

You have an opportunity to gamble on the toss of a coin. If your choice is correct you win £30, but if it is wrong you lose £50. Initially you think it is equally likely that the coin is two-headed, two-tailed or fair. If you are risk neutral, how much should you be willing to pay to observe a size 1 sample, i.e. one throw?

After the information given about the outcome of that throw, you have three choices: (1) do not bet; (2) bet on heads and (3) bet on tails. The message service tells you either head or tail. If the message is $m$ = head, then you can calculate how the probabilities (the posterior probabilities) have altered.

Let's deal with each column in Table 5.1 in turn. At first, you don't know whether the coin is two-headed, fair, or two-tailed. There are three options, all equally likely, and therefore a 1/3 chance that any of those is correct. If the coin is genuinely two-headed, the likelihood of getting the message $m$ = head, is definitely going to happen. If the coin is fair, the likelihood of getting the message $m$ = head is only 1/2 as the coin could turn up a head or a tail. If the coin is two-tailed there is a zero chance of getting the message $m$ = head. The joint probability is the prior probability times the likelihood of message $m$, and the sum of the joint

**Table 5.1**  Probability matrices for coin tossing

| State of the world | Prior probability (P) | Likelihood of message m (q) | Joint probability (P × q) | Posterior probability (Joint probability/k) |
|---|---|---|---|---|
| Coin is two-headed | 1/3 | 1 | 1/3 | 2/3 |
| Coin is fair | 1/3 | 1/2 | 1/6 | 1/3 |
| Coin is two-tailed | 1/3 | 0 | 0 | 0 |
|  | 1 |  | k = 1/3 + 1/6 = 1/2 | 1 |

probabilities is summed as $k$. The posterior probability expresses the changed circumstances, now that we have message $m$ = head. As for the joint probability, the overall chance of heads, given message $m$ = head, is:

$1 \times 2/3 + 0.5 \times 1/3 = 5/6$

(i.e. multiplying the posterior probability by the likelihood of message).

The best action is therefore now to bet on heads.

The expected utility of the message is calculated, loosely speaking, as the chances of gaining so many pounds less the chances of losing so many pounds:

- The chances of heads is 5/6 multiplied by £30 (gain) = 150/6.

- The chances of tails is 1/6 multiplied by £50 (loss) = 50/6.

- The expected gain in utility from the message is 150/6 − 100/6 = 50/6 = £16.67.

This is not the sort of calculation you are likely to be making on a daily basis but the principles generally apply to many situations. Informational decision analysis requires calculation of the following:

- *Prior probability matrix* representing the set of probability beliefs, for various possible states and actions associated with those states.

- *Likelihood matrix* representing the set of conditional probabilities of any message ($m$) given any state ($s$).

- *Joint probability matrix* representing the set of probabilities of states ($s$) and messages ($m$).

- *Potential posterior matrix* representing the set of conditional probabilities of each state given any message ($m$).

## Belief revision

The example of disease diagnosis will be used to show how prior probabilities are converted into posterior probabilities in the process of belief revision. The example lends itself well because of the possibility that a test will come up negative when the disease is actually present, or positive when the disease is actually absent.

In this example, a patient may or may not have a particular disease (states are disease absent and disease present). A test will help to reveal whether or not the disease is present or absent. There is a 90 per cent chance of the test proving positive if the disease really is present, and an 80 per cent of the test proving negative is the disease really is absent. The likelihood matrix is shown in Table 5.2.

The first row shows that if the patient actually has the disease, there is 90 per cent chance of the test actually showing up as positive, but a 10 per cent of the test wrongly showing a negative result. As the test being negative or positive are the only two possible conditions, 0.9 + 0.1 = 1.0.

Probabilities can be attached to the states (disease absent, disease present). Assume that there is prior probability of the disease being present of 0.6 (and hence 0.4 of it being absent). The joint probability matrix can therefore be represented by Table 5.3.

**Table 5.2**   Likelihood matrix for diagnostic test

| States | Message = test negative | Message = test positive | |
|---|---|---|---|
| Disease present | 0.1 | 0.9 | 1.0 |
| Disease absent | 0.8 | 0.2 | 1.0 |

**Table 5.3**    Joint probability matrix for diagnostic test

| States | Message = test negative | Message = test positive | |
|---|---|---|---|
| Disease present | 0.06 (0.6 × 0.1) | 0.54 (0.6 × 0.9) | 0.6 |
| Disease absent | 0.32 (0.4 × 0.8) | 0.08 (0.4 × 0.2) | 0.4 |
| | 0.38 | 0.62 | 1.0 |

**Table 5.4**    Potential posterior matrix for diagnostic test

| States | Message = test negative | Message = test positive |
|---|---|---|
| Disease present | 0.16 (0.06/0.38) | 0.87 (0.54/0.62) |
| Disease absent | 0.84 (0.32/0.38) | 0.13 (0.08/0.62) |
| | 1.0 | 1.0 |

This table reminds us that we can get a negative test result if the disease is present (this is less likely, but the possibility exists). Similarly we can get a positive test even if the disease is absent.

The potential posterior matrix can be calculated as per Table 5.4.

The posterior probability that should be attached to a particular state, after receiving message $m$, is equal to the prior probability, multiplied by the likelihood of message $m$, and then divided by a factor which is the overall probability of receiving message $m$. The far right-hand column shows the posterior probability, the chances that the patient actually has the disease, if the test is positive. This is the figure taken from the previous matrix (the prior probability times the likelihood of receiving a positive test result), divided by the overall probability of receiving a positive test result (which is calculated in the previous matrix as 0.62). The important thing to note is that there is still a chance that the patient may not have the disease, even if the test is positive.

This is known as the Bayesian revision of probabilities, after Thomas Bayes the clergyman who first introduced this equation.

The way of writing the equation is:

$$P(C/pos) = \frac{P(pos/C)P(C)}{P(pos/P(C) + P(pos/notC)P(notC)}$$

In words, this means (on the left-hand side) the probability of C (having the disease), given a positive result is calculated as the probability of getting a positive result if the disease is present times the probability of getting the disease anyway, divided by the probability of getting a positive result (whether the disease is present or whether the disease is absent).

We can substitute as follows:

- for *patient with disease* absent or present, read *supplier* who is healthy financially or not;

- for test, read *grapevine gossip.*

We then keep the probabilities as they were, in other words there is a 90 per cent chance of the gossip being right if the message is that the supplier is not financially viable, and an 80 per cent chance of the supplier being financially healthy and viable if the grapevine gossip says the supplier is financially healthy. There is a prior probability of the supplier being financially unhealthy, which you have assessed as 0.6. If the subsequent grapevine gossip says the supplier's finances are not good, then the posterior probability is greater than 0.6 (it has increased, as shown in the table, to 0.87).

| Table 5.5 | Potential posterior matrix for assessing supplier's financial health | |
|---|---|---|

| States of supplier | Gossip message = supplier viable | Gossip message = supplier not viable |
|---|---|---|
| Not financially healthy | 0.16 (0.06/0.38) | 0.87 (0.54/0.62) |
| Financially healthy | 0.84 (0.32/0.38) | 0.13 (0.08/0.62) |
| | 1.0 | 1.0 |

Our posterior probability matrix needs to be relabelled (Table 5.5).

This is not a perfect example as the grapevine does not have the scientific accuracy of an actual test, but you can see how the process of belief revision may be modelled. On balance we were slightly concerned before we heard the news, after the news, we have every right to be more concerned. In this case the gossip is credible but the matrix would look different if we had little reason to believe our source of gossip.

Assuming the grapevine gossip is not really very credible, you can estimate that the gossip will rightly identify problems 70 per cent of the time, and that only 25 per cent of the time will the gossip be totally accurate that suppliers are in a genuinely healthy financial state. The joint probability matrix (Table 5.6) looks very different this time, particularly as we know that only 1 per cent of suppliers have financial problems. If the gossip says the supplier is in difficulties, should we believe them?

The probability of the supplier not being viable if the gossip indicates that the supplier is not financially healthy is calculated in the potential posterior matrix, using the same system as before (Table 5.7).

In fact, although the source of gossip may be 70 per cent accurate in identifying problems, the chances that the

| Table 5.6 | Joint probability matrix for less reliable gossip |
| | |

| States of supplier | Gossip message = supplier viable | Gossip message = supplier not viable | |
| --- | --- | --- | --- |
| Not financially healthy | 0.003 (0.01 × 0.3) | 0.007 (0.01 × 0.7) | 0.01 |
| Financially healthy | 0.2475 (0.99 × 0.25) | 0.7425 (0.99 × 0.75) | 0.99 |
| | 0.2505 | 0.7495 | 1.00 |

| Table 5.7 | Potential posterior matrix with less reliable gossip, healthier suppliers |
| | |

| States of supplier | Gossip message = supplier viable | Gossip message = supplier not viable |
| --- | --- | --- |
| Not financially healthy | 0.013 (0.003/0.2505) | 0.0093 (0.007/0.7495) |
| Financially healthy | 0.988 (0.2475/0.2505) | 0.991 (0.7425/0.7495) |
| | 1.0 | 1.0 |

supplier is really not financially healthy, given the news that they are, is nothing to worry about. The potential posterior matrix indicates that they are still 99 per cent likely to be financially healthy, despite the gossip message to the contrary. The conclusion is to stay calm and think of the 'base rate', the chances of the supplier being unhealthy in the first place.

# Problems with Bayesian inference

In real life, the posterior probabilities are not calculated neatly but estimated approximately in people's heads. Unfortunately, people are not very good at accounting for all parts of the Bayesian equation. Given a situation where people are given a description of someone's characteristics, judgments of the probability that the person belongs to a particular category (of occupation, for example) is done solely on the match between the information known about that person and the characteristic member of the category. We rarely think about the likelihood of finding anyone of that category. If the description fits the stereotype (glasses, cardigan and sensible shoes = librarian) then the person is judged likely to be a librarian, even if information is supplied that suggests that the chances of finding a librarian among that group of people is quite unlikely.

In the above example, we might also be more likely to believe that the supplier was financially unhealthy if we had been reading about similar bankruptcies recently. You would have a recent memory of a similar event that makes the story more credible.

## Saving face and minimising regret

One of the dilemmas facing managers is the difficulty of making decisions that depend on what might happen in the future. We do not always know what the level of demand might be, but we do not want to seem unprepared to meet demands for services. On the other hand, we do not want to invest in a service that is hardly used.

Imagine we could plan for service provision for a digital library service support in one of three ways: do it in-house, contract out or join a shared services consortium. The in-house provision has the advantage of skilling up the existing workforce, but there are the training and investment costs, and the problems of redeploying expensive staff if the demand for the service is not high. Contracting out is more flexible, does not require additional in-house staff costs, but there is always the worry that a small business might fold, or not be able to scale up when required, should the service demand increase. Joining a shared services consortium is not as flexible as contracting out, but may offer a better quality of service and more job satisfaction to existing staff in the future. If we estimate the human resource benefits or losses on a scale from +20 (very satisfied, fully deployed staff) to –20 (staff trained unnecessarily, loss on training investment, unhappy staff) then we could set a benefit matrix as shown (Table 5.8). If future service demand is low, then in-house training will not be worthwhile, and at medium levels the costs may just level out. For low levels of demand, it is worth contracting, as it is much less hassle to existing staff. At high levels of demand, however, there is the likelihood that staff will have to cope with pressures of demands that they cannot meet, and that the contractor cannot satisfy either. With shared services, the start-up costs have to be met, but at medium levels of demand, the staff training needs can be easily accommodated.

If you were a very optimistic manager you might want to go for the decision that offered the greatest rewards, the maximum possible benefits for staff skills and job satisfaction. If you study the matrix you will find that the decision to train in-house offers the highest of all possible payoffs. The maximum possible benefit for contracting out or shared services is only 15.

**Table 5.8**   Benefit matrix

| Decision | Future demand | | |
|---|---|---|---|
| | Low | Medium | High |
| In-house | –20 | 0 | 20 |
| Contract out | 15 | 5 | –15 |
| Shared services | –10 | 10 | 15 |

This option could be very risky if future demand is low. You might prefer to play safe, considering the worst possible outcome of each decision, then choosing the decision that offered the best of these. This is the saving face decision. If the worst came to the worst, training in-house could leave you with a hole of –20, contracting out could leave you worse off by –15, and shared services by –10. Clearly, the safest option, if you wanted to play safe, and were pessimistic by nature, would be to select the shared services option. That leaves you with a deficit, but not as much as the other two options.

The third approach is to calculate the opportunity loss or regret. For each of the possible levels of demand, there is one decision that is best in those circumstances, but if you look at the medium level of demand column you will notice that there is not much difference in benefits between choosing contracting out or shared services. In other words, while shared services is the best option, you are not going to regret a decision to contract out by that much. On the other hand, if you chose to train in-house, and the future demand turned out to be low, there is a much bigger gap between contracting out and training in-house, and you really would be kicking yourself if you had chosen the latter.

The way of calculating the regret matrix[1] is to study each column in turn. The maximum possible benefit in the low demand situation is 15. For the low demand situation, the best option is to contract out, and you would have no regrets if you had made that decision (15 – 15). If you had decided to train in-house, the maximum possible benefit in that setting is 15, but your benefit level is –20. Your regret would be 15 – (–20) = 35 (if this calculation troubles you, think of the difference along the line between –20 and +20). If you had chosen shared services, your regret would be 15 – (–10) = 25.

**Table 5.9**    Regret matrix

|  | Future demand | | | |
|---|---|---|---|---|
|  | Low | Medium | High | Maximum regret |
| In-house | 35 | 10 | 0 | 35 |
| Contract out | 0 | 5 | 35 | 35 |
| Shared services | 25 | 0 | 5 | 25 |

Table 5.9 presents the regret matrix if you do similar calculations for the other columns.

The lowest maximum regret in this situation is to choose to work in a shared services consortium (25), as both in-house training and contracting out would incur greater levels of regret.

Taking all of the possible approaches to deciding on the best option, shared services came out top in two. If you could be sure that the chances of low, medium or high levels of demand were equally likely, then opting for shared services may be the best approach. Working through these types of matrices does not give you a guarantee against disaster but the process may help you to think through the consequences of managerial decisions.

# Assessing probabilities and expected values

If you had discussed the decision on in-house training, contracting out, or joining a shared services consortium with your senior management team they might come up with different views on the wisdom of the decision.

Miss Adventurous claims that demand will be high and that going for shared services will be missing a valuable opportunity to develop in-house skills and going one better than library services that are usually benchmarked against you.

Mr Cautious disputes this, claiming that demand is likely to be low for the next three years at least. With restructuring on the horizon anyway, why tie ourselves to a consortium agreement that may need to be renegotiated within the next three years?

Mrs Cooperative likes the idea of the shared services model, but is a bit worried about the idea of restructuring. She thinks the level of demand is mostly likely to be medium.

And so it goes on. After taking a vote among the ten members of the team, five members voted for low demand being the most likely outcome, four for the medium demand, and only one, Miss Adventurous thought that demand would most likely be high.

The consensus view of the meeting was that the probability of low demand was assessed at 5/10 (0.5), medium demand at 4/10 (0.4) and high demand 1/10 (0.1). Alternatively, the team could have used independent market research estimates of the probability of each of the demand scenarios, or based their estimates on previous experience with similar services. The probabilities add up to one (all possible outcomes add to one: tossing a coin: 1/2 heads, 1/2 tails, adding to one; rolling a die, 1/6 chance of a two, three, four, five, six or one, and six times 1/6 equals one).

The regret matrices were based on the assumption that each demand outcome was equally likely, but now we need to take the different probabilities into account. We can calculate an expected value for each decision option, although we need to remember that this value is only a way of comparing the decisions.

The expected value for in-house training is:

$$0.5 \times -20 + 0.4 \times 0 + 0.1 \times 20 = -10 + 0 + 2 = -8$$

The expected value for contracting out is:

$$0.5 \times 15 + 0.4 \times 5 + 0.1 \times -15 = 7.5 + 2 - 1.5 = 8$$

The expected value for shared services is:

$$0.5 \times -10 + 0.4 \times 10 + 0.1 \times 15 = -5 + 4 + 1.5 = 0.5$$

Given the greater probability of a low demand, the expected value comparison indicates the highest expected value is associated with the contracting out option. That is not the value that you would obtain, of course, as we have already estimated that you would actually receive 15, 5, or −15 depending on whether demand is low, medium or high. The expected value is simply a way of taking into account the probability of other outcomes.

## Presentation of risk

Presenting this information to the Board should be done carefully. Psychological research[2,3] on the framing of risk indicates that people will react differently to the same risk, depending on whether it is presented as a gain or a loss. In addition, some people are more risk averse than other people.

If your senior management team had decided that the choice came down to shared services or contracting out, you need to be careful about the way you present the likely gains or losses. You could, quite truthfully, point out that that there was:

- a 10 per cent chance of obtaining a value of 20 with in-house training (with high demand);

- a 50 per cent chance of obtaining 15 for contracting out (with low demand);

- a 40 per cent chance of obtaining 10 for shared services (with medium demand).

This is not the whole story, but you are presenting the possible gains, the good news. It is very likely that the Board, presented with this information would choose contracting out, as that seems to offer the surer gains. However, the Board should query possible losses, and the risks of leaving all of you feeling very uncomfortable about the decision. Truthfully, you could present them with the following information. You would point out that there is:

- a 10 per cent chance of incurring a loss of 15 with contracting out (with high demand);

- a 50 per cent of incurring a loss of 20 with in-house training, or a loss of 10 on shared services (with low demand);

- a 40 per cent chance of just breaking even on in-house training (with medium demand).

Given the information about the possible losses, an extremely risk averse Board might decide that in-house training was definitely not a preferred option. The debate over the merits of contracting out versus shared services might keep the Board engaged in discussion for a long time. There's a 10 per cent chance of losing 15, versus a 50 per cent chance of losing 10 – a difficult decision for some Boards, indeed.

# Final words on presentation of risk and benefits

The chapter has examined how we can work through how our beliefs should be revised, given new information, and how we can structure decisions made about planning for future services. We can use regret matrices, and we may also calculate 'expected values' although it must be stressed that these are a theoretical value. The previous paragraph has stressed the way information about gains and losses can skew our decision making. National lotteries would not be very successful if we were all rational decision makers. The chances of making huge gains are very tempting, and we forget that the chances of winning are miniscule. We also have an unfortunate habit, well documented by psychological studies of decision making, of seeking information that confirms our existing beliefs, and ignoring or downplaying information that appears to contradict our beliefs.

Another financial trick is the presentation of increase in benefits or risk. Should we be worried if the new member of staff makes 50 per cent more cataloguing errors of type A than the experienced member of staff she replaced?

That sounds rather scary, but it depends on the way we present the information and how likely it is that such errors will occur in any case. We could have situation A:

- Experienced cataloguer, annual errors of type A: 200.

- New staff, annual errors of type A: 300.

- Relative percentage increase in cataloguing errors of type A

$$= \frac{300 - 200}{200} \times 100 = 50\%.$$

- Total number of cataloguing errors (all types) estimated to be 400 annually for the experienced cataloguer.

- New total estimated to be $400 + 100 = 500$ (all attributable to type A errors).
- Relative change in total cataloguing errors overall

$$= \frac{500 - 400}{400} \times 100 = 25\%.$$

Or we could have Situation B:

- Experienced cataloguer, annual errors of type A: 10.
- New staff, annual errors of type A: 15.
- Relative percentage increase in cataloguing errors, type A

$$= \frac{15 - 10}{10} \times 100 = 50\%.$$

- Total number of cataloguing errors (all types) estimated to be 400 annually for the experienced cataloguer.
- New total estimated to be $400 + 5 = 405$ (all attributable to type A errors).
- Relative change in total cataloguing errors overall

$$= \frac{405 - 400}{400} \times 100 = 1.25\%.$$

Moral: always ask percentage of what!

## Notes

1. See, for example, Wisniewski (1997) Chapter 6.
2. See, for example, Tversky and Kahnemann (1974, 1981).
3. See also Paulos (1988: 82–94).

# Forecasting and simulation

This chapter explains some of the ways in which we can predict values of interest to us in the future. We may use linear regression to examine whether previous trends can be used to predict what will happen in the next year. Logistic regression examines the likelihood of some events occurring, given what we already know about some aspects of service usage. Rationing is almost inevitable given the fact that demand for some services is always going to exceed resources available to the manager. The queues at the service desk are a very obvious result of this, and queuing theory examines how to estimate whether it will be worthwhile opening up another service-desk point, or whether we would have staff idle for a considerable proportion of their time. Simulation is another way of envisaging what might happen if we alter the organisation of our services – would we be able to cope with the bottlenecks in demand any better?

## Linear regression

If we believe that there is some direct relationship between two variables, we may use that relationship to help us predict what might happen in the future.[1] For example, as funders of library services in an area of the country, we might notice a relationship between the funding levels and the number of journal subscriptions a particular library service has. If we plotted, for each library service, the number of subscriptions against the funding level, we might end up with a diagram such as Figure 6.1.

In this scatter diagram, you should observe that the slope of the line is upward, to the right. It seems that the more funding that is provided, the greater the number of journal subscriptions. You will also notice that even if there are very few journal subscriptions, the library still requires a basic level of funding, as there are other services to be provided. There is one outlier, one library that does not seem to fit the general pattern. When we take a closer look at this, we might find that it has a different funding model from the other libraries. For our purposes here, we can ignore this outlier in the calculations (see Figure 6.2), but the reasons for the outliers must always be investigated before ignoring them.

When we are faced with a scatter diagram we would usually prefer to have one line that was some expression of

**Figure 6.1** Basic scatter diagram of funding against number of journal subscriptions

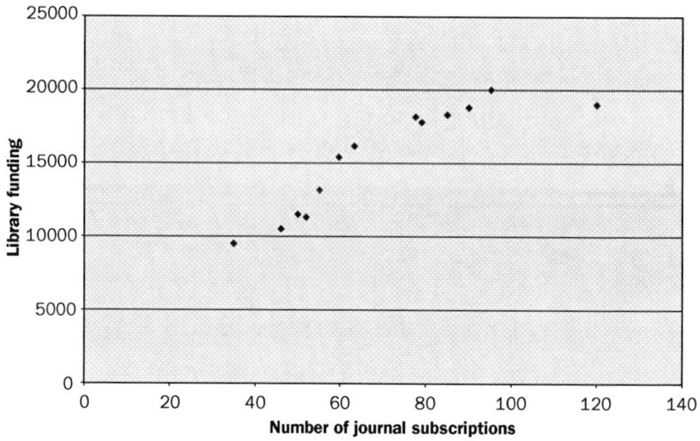

**Figure 6.2**    Scatter diagram of funding against journal subscriptions, with linear regression line

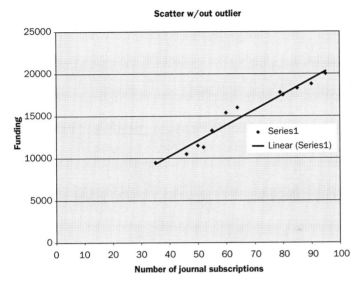

the happy compromise between the extremes, the best fit line as it is called. In fact, the mathematical way of finding the best fit is to find the line that is placed so that the sum of the residuals, squared, is at a minimum. A residual is the distance from one of the points to the line.

The equation for the line is:

$$\text{funding } (Y) = 2{,}937 + 182.88 \times \text{the number of journal subscriptions } (x).$$

It is important to be aware of the assumptions we are making in this linear regression analysis. The line fits the data quite well, and a calculation of the proportion of the variability of $Y$ that can be explained by its relationship with $x$ is in fact 95%. In other words, nearly all the differences in funding between these library services can be explained by the number of journal subscriptions they have.

**Figure 6.3**   Residuals scatter diagram

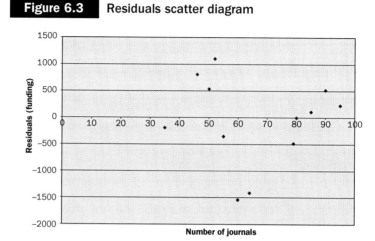

We should also check whether the residuals produce a random scatter of points when plotted against $x$ (number of journals), as they do in Figure 6.3. The residuals should also have the same variability for all the fitted values (predicted, calculated values) of $Y$.

Once we are happy that we do have a regression line that is genuinely a good fit, we can use the equation to predict, for example, what the expected funding levels for a library with 70 journal subscriptions would be:

$$\text{funding (expected)} = 2{,}937 + 182.88 \times 70$$
$$= £15{,}378.60$$

If the scatter increased or decreased as $Y$ increased, then we would have to suspect that the variation in funding was greater or lesser than we had assumed for different numbers of journal subscriptions. That would make drawing the line of best fit more problematic.

As mentioned in the introduction, some of the earlier use of quantitative methods in library science has not developed greatly since the 1980s, and finding examples meant hunting back in the literature. It is important to remember that much of this work was done prior to the development of Microsoft Excel spreadsheets. Such spreadsheets and their built-in functions make life easier in some respects, but they may also be limiting, as most people will only use what is there in the package. An early study that analysed and tried to estimate forecast demand for interlibrary loan services in a regional network examined various methods of trying to separate the pattern from the randomness of the data. Kang and Rouse (1980) discuss how to deal with seasonal variations, and present three different forecasting methods (Box-Jenkins time series, adaptive filtering and linear regression). The article is probably now more interesting for the discussion of some of the assumptions behind the techniques and the adaptations that might be made. For example, the 'fading memory' regression technique is simply a way of weighting the more recent data more heavily than the data from farther back in time. This makes sense in many situations.

## Multiple linear regression

In the above examples, there have only been $x$ and $Y$ to consider, only one explanatory variable to explain the regression of $Y$ on $x$. We may, however, suspect that there are several explanatory variables. The multiple linear regression equation is:

$$Y = a + b_1 x_1 + b_2 x_2 + \dots b_k x_k$$

Just as in the simple linear regression equation, $a$ is the intercept, the value of $Y$ when all the $x$ values are zero. The

$b$ values are the estimated partial regression coefficients, $b_1$ represents the effect of $x_1$ on $Y$ that is independent of the other $x$ values. Not surprisingly, multiple regression analysis is something that is performed on a computer. Just as in simple linear regression analysis, it is important to check for the goodness of fit, and that the linear regression equation is suitable. We also need to check that the variation we would expect when measuring and estimating $x$ for various values of $Y$ is more or less the same – that we have a well behaved equation. If you can imagine that the most likely range of variation is contained within two lines above and below our line of best fit, we want to be sure that there are no points along the line where we would find bulges or that our parallel lines diverge.

## Relationships that are not linear

When we plot a scatter diagram it may be obvious that there is some sort of curve, but that it is not a straight line. In some circumstances, it is possible to use the shape of the curve to deduce what type of relationship exists between $Y$ and $x$. Often, logarithmic transformations are used to develop a non-linear model.

## Estimating trends

It may be possible to estimate a trend using linear regression techniques, but remember that we need to be fairly sure that things are not going to change significantly. Linear regression is useful when you are trying to spot a likely upward or downward trend among considerable seasonal variation. Figure 6.4 shows the seasonal trends in requests;

**Figure 6.4**    Seasonal variation in requests

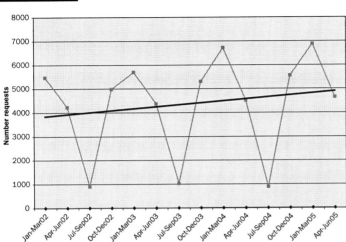

the manager would like to know whether there is a general trend upwards or downwards. It seems to be generally upwards, but by how much? Would it be possible to estimate the likely number of requests for the last two quarters of 2005?

The linear regression equation indicates an upward trend, but it is obvious that predicting the results for the next quarter may be difficult. Another warning sign is that the test of goodness of fit is not very comforting news – an $R^2$ value of 0.03 tells you that only 3 per cent of the variability can be explained. If you compare the results for each quarter, i.e. January–March and July–September, you will find that while most of the quarters show an increase year on year, the figures for the July–September quarters may be in decline. Figure 6.5 illustrates the results of linear regression analysis on just the July–September quarter figures alone.

The equation on the chart is the linear regression equation, and the minus sign tells you that $Y$ is decreasing as

**Figure 6.5** Summer seasonal trends

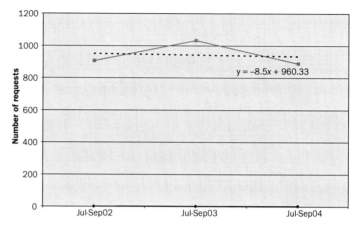

July-September quarter trends

$y = -8.5x + 960.33$

$x$ is increasing. If you used this subset of the data to calculate the figure for July–September 2005, you would estimate a total of 934 requests for the quarter. On the other hand, if you used the general linear regression trendline, obtained from the entire series, with all quarters included, the trend figure would be 4,985. The lesson is that you need to look carefully at the figures and charts first before applying the statistical manipulations. In this situation, something is causing an apparent huge and declining drop in the number of requests over the summer periods, although other factors may be contributing to the general rise in the number of requests. If you did a similar calculation for the October–December quarters, you would estimate the October–December 2005 figure as 5,866. Using the general trendline your estimate would be lower, at 5,066. The general trendline will overestimate the low quarters, and underestimate the high quarters. Although the yearly totals may be approximately correct, this is not much help to you when planning how many staff you need dealing with requests over the summer period when many may be requesting leave. Using the general trend without careful inspection of the patterns within the trends may leave you with very unmotivated staff with time on their hands during the summer.

The moral is to use the scatter diagram first and eyeball this before becoming distracted by lines or columns. The scatter diagram for the requests (Figure 6.6) shows very clearly that there are two sets of behaviour involved. What happens in the 3rd quarter, the 7th quarter and the 11th quarter, all the July–September quarters, is different from what seems to be happening in the other quarters.

Another way of dealing with this problem is to quantify the trend and seasonal components. The trend is the long-term underlying movement. The seasonal components reflect, as shown in the example, the repeating pattern, in this case for lower numbers of requests over the summer period.

The long-term trend can be estimated using the principles of moving averages, in other words, averaging out the seasonal components. The diagram produced, averaging

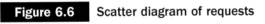

**Figure 6.6**   Scatter diagram of requests

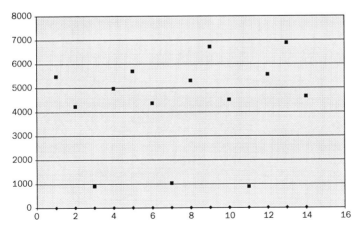

over four quarters (which is why the trend only starts after January 2003) shows a small upward trend, with a sharper increase at the end of 2003.

Figure 6.7 shows the predicted values, using the values calculated separately for the changes in the summer quarters (possibly a slight decline) and the changes in the

**Figure 6.7** Predicted values (based on separate quarterly estimates)

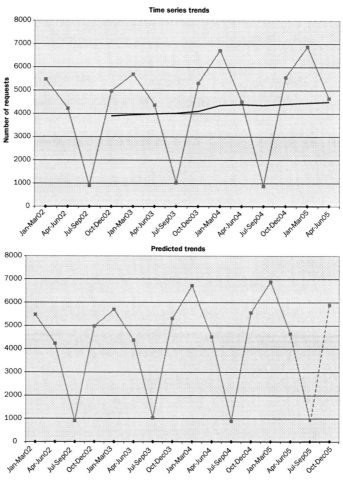

October–December quarter (generally rising). The values from April–June 2005 are predicted values (indicated by the non-solid lines for the associated data points). Figure 6.8 shows the same graph, but with the moving average trendline added, for averaging over the two previous quarters. Figure 6.9 shows the same graph but with the moving average trendline for the previous four quarters added. This is much closer to a straight line, as the seasonality components are effectively ironed out over the four quarters in this calculation. When the moving average is calculated on the basis of the previous two quarters, the seasonal variation is apparent. On average, therefore, the trend over the year is for a slight rise, but this conceals the fact that there are seasonal variations.

An example of trend analysis is the research conducted by Loria and Arroyo (2005) examining the increasing dominance of the English language in MEDLINE. Linear regression analysis was used to examine the relationships

**Figure 6.8** Predicted trends with moving average trendline (over two quarters)

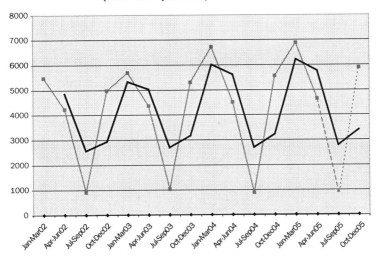

**Figure 6.9** Predicted trends with two types of moving average trendlines

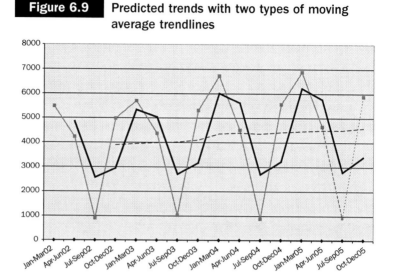

between yearly number of papers and year of publication. The results showed that while the total number of articles is increasing steadily, year on year, the non-English papers coming from the non-anglophone countries (anglophone countries considered as Australia, Canada, England, Ireland, New Zealand, Scotland, USA and Wales) are actually declining. This is attributed partly to the deselection of journals from non-anglophone countries (and some publishers have reacted by increasing their output in English, anglicising their title and changing the publication site to an anglophone country site).

## Logistic regression

If you wanted to profile the interlibrary loan behaviour of your users, and wanted to test whether the presence or absence of some characteristics made a difference to their interlibrary loan behaviour, you would use a similar, but different type of regression equation, called logistic regression. As the user is, for example, either a scientist or not, a postgraduate or not, these variables are either present or absent, and we need to use a different type of transformation. This type of calculation requires a computer package and preferably some statistical advice. Like the linear regression equations, you need to look for the 'goodness of fit', a test that indicates that the approach is realistic, as well as the information that tells you whether or not any of your variables are associated with the interlibrary loan behaviour. If the calculated odds ratio for postgraduate was 4.3, this would indicate that postgraduates were over four times as likely as non-postgraduates to be interlibrary loan users. However, you also need to check the confidence limits on the odds ratio as well (and the $p$-values) to be sure that being a postgraduate (or not) is a significant factor in determining interlibrary loan behaviour. In other words, for planning service provision, you need to be sure whether or not an influx of hundred or so more postgraduates will have a significant effect on demand. The confidence limits can be calculated, and mean something quite distinct statistically, as do the $p$-values or probability values, but the most important thing to remember is that the confidence limits express, approximately, how sure you are that 4.3 is about the right value.

Logistic regression is used in studies of transaction logs of Web usage. If, for example, you are interested in devising a personalisation system, it is important to identify the types of resources and links that particular types of user are most likely to use. Bracke (2004) used multinomial logistic regression to examine whether the type of resource affected the way the users navigated to the gateway page for that resource, and what the differences in navigation patterns were for 'average users', on-campus users and off-campus users. The analysis illustrated how the on-campus and off-campus users navigated to journals, textbooks and databases, and the patterns varied, depending on whether the user was off-campus or on-campus, and what the resource required was. The analysis also examined the different patterns of searching for the resources, such as going through the library catalogue (not at all popular in this case), searching for a known item (more popular for journals than for other types of resource), or browsing (generally the preferred route). For example, the probability of an on-campus user browsing to a journal was 0.37, searching for a known item was 0.51, and going through the catalogue was 0.12.

With the amount of information available on the Web, creating an intelligent agent to do the sifting for the type of pages that your users want to see, seems an eminently sensible procedure. Scott Nicholson (2003) describes research that compared various ways of estimating the likelihood that a web page would be the type of page required, in this case, scholarly research web pages. Such pages should be written by students or faculty, produced by a non-profit making research institution or published in a scholarly peer-reviewed journal. The selection criteria were gathered from a literature review and then subjected to a review panel of 42 librarians, who ranked the criteria, and made further suggestions. A final list included criteria for the author, content, organisation, producer/medium criteria and external criteria.

One of the models tested was a logistic regression model, a stepwise logistic regression model that selected the most important subset of the variables to create a useful prediction model. Some of the criteria are intuitive, and might be those a human expert would use in judgment, such as 'presence of labelled bibliography' or 'academic URL', but others are less obvious (such as the number of misspelled words according to Doctor HTML). Nevertheless their presence or absence helps to predict whether or not a particular web page is a scholarly research work. Other models tested included a classification tree, which ended up with 13 criteria, and used some 'if this, then that' reasoning to decide whether or not a page was scholarly research.

## Queues and rationing

Queues and service desks are a combination known to most service managers. The perpetual queue ensures that staff are always kept occupied, but a long queue means dissatisfied and impatient customers, as well as staff who feel stressed. Is there a happy medium, and how can that be estimated?

First, we need to study the way queues are forming. Are the customers arriving on a fairly regular basis, or does the arrival rate vary according to the time of day? In other words, would it be a fair estimate to say that, on average, 40 customers (or another number of customers) would arrive per hour? Note that we are not saying that every customer will arrive neatly one minute and so many seconds after the preceding customer, simply that on average the arrival rate is 40 customers per hour. However, experience may suggest that the arrival rates vary according to the time of day. We may have problems with batch arrivals, such as a large class of schoolchildren arriving, or a lunchtime or early evening session, when more people will have time to drop into the library to borrow a book. Another factor to consider is whether our customers are acting independently – is the time of arrival of one customer dependent on the arrival time of another? This question is not aimed at understanding customer behaviour, but merely to allow us to determine what type of equations we can safely use to analyse our queues and the number of service points we need. Customers would like to know how long, on average, they will have to wait, and what the chances are that they might have to wait longer than they deem acceptable – will they miss their bus or will the parking time be exceeded?

We can use simple queuing theory if:

- there is a simple queue – one queue;
- the average arrival rate is known or can be measured;
- the average service rate (rate at which staff serve customers) is known or can be measured;
- there are no batch arrivals;
- there are no time dependent patterns (or that these can be excluded).

Note that we do not assume that the arrival rate will always be the same or that the service rate is constant. These will vary, but we can assume that the distribution will be of a particular standard type (usually Poisson). Much of the time in library and information services we are dealing with rare events that occur randomly and independently. Most of the time a book on the library shelf stays there, but when books are borrowed, the decisions made by Mr Jones of 27 Acacia Avenue and Miss Smith of 16 Maple Grove when they borrow books are made independently. What Mr Jones chooses is not affected by the decision of Miss Smith. The number of loans per day is often likely to follow a Poisson distribution. There are of course circumstances when we would not expect to find a Poisson distribution as our users are not acting independently. An example of this would be the loan distribution when a class of students is given an assignment and a reading list at the same time.

If, for the queuing system and our service points, we now assume that the system has reached a steady state, some type of equilibrium, then we can estimate the traffic intensity, the degree of congestion, the probability that a customer has to wait, and the probability that the staff serving are not doing anything.

If the average (mean) customer arrival rate is five customers per minute, and the average (mean) service rate is four customers per minute, then it is easy to appreciate that queues are likely to build up, with a traffic intensity of arrival rate/departure rate of 5/4 = 1.25. In reality we would try to handle the situation better and a more realistic situation might be an arrival rate of three per minute and a service rate of four per minute. That gives a traffic intensity of 3/4 = 0.75. The nearer to one (or more than one) the traffic intensity is, the more likely that there will be queues. The nearer to zero the traffic intensity is, the more likely that there will be no queue for an individual customer approaching the service desk.

For a simple situation in which customers are arriving at a mean rate of one every five minutes, and the service rate is one customer every two minutes, with one queue and one service point, then:

arrival rate = 0.20 (1/5) per minute
service rate = 0.5 (1/2) per minute
traffic intensity = arrival rate/service rate = 0.20/0.50
= 0.4

To calculate the probability of no customer, one customer, two customers or more than two customers arriving within a time period, we use the Poisson distribution:

$$P(X = x) = e^{-\lambda}\frac{\lambda^x}{x!}$$

Some explanations of the terms in this equation are required. Conventionally the 'mean' in the Poisson distribution is denoted by the lambda symbol $\lambda$.

The $e$ symbol stands for exponential, a type of mathematical function that turns up in many types of mathematical calculations, but we do not need to explain it further here.

The factorial sign ! is a short way of indicating a string of numbers to be multiplied together. $4! = 4 \times 3 \times 2$. Remember that $0! = 1$ (illogical maybe but that is how it is).

For ease of calculation we shall use a five-minute period, so that $\lambda = 1$.

$P(0) = 1^0 \times e^{-1}/0! = 1 \times 0.3679/1 = 0.3679$
(0.4 approximately) (nobody arrives)

$P(1) = 1^1 \times e^{-1}/1! = 1 \times 0.3679/1 = 0.3679$
(0.4 approximately) (one person arrives)

$P(2) = 1^2 \times e^{-2}/2! = 1 \times 0.1353/2 = 0.07$
(0.1 approximately) (two people arrive)

We could go on to calculate the chances of three, four, and five people arriving, but the probabilities will decrease, as you might guess.

In this situation, the chances that your service desk is idle is quite high, but you may be concerned that you want to minimise the queues. If you decided to bring another service point into operation, but still with one queue, what difference would you make to customer service?

arrival rate = 0.20, as before.
service rate = 1.0, double the service rate before.
traffic intensity = arrival rate/service rate = 0.2.

The probability that a customer will wait in those circumstances can be estimated by thinking of the occasions when they will not have to wait, and subtracting that from one (as all probabilities have to add to one – think of rolling a die, where there is 1/6 chance of a six, 1/6 chance of a two and so on, but six times 1/6 = 1).

In our five-minute period, and assuming that no customer is being served at the start of the period, no customers $P(0)$, one customer $P(1)$, or two customers $P(2)$ may appear. If we have two service points, there will not be a queue under those circumstances, as everyone can be served.

$$\text{Probability of waiting} = 1-(0.4 + 0.4 + 0.07) = 1-0.87$$
$$= 0.13$$

Conversely, of course, the probability that your service desk staff are idle is quite high.

Using a software package we can also calculate:

■ The effect of increasing productivity of the staff on the service point.

■ Costs of different queuing options. For example, it is quite likely that customers seeing a long queue will go away without being served. That may cost us photocopier income.

■ The effect of different types of queue orderliness. Normally we would expect that first come, first served should prevail but we might decide to assess whether we need to separate the queues into long queries and short queries.

## Simulating the workflow

For many situations the assumption that there is a steady state is a little tenuous, and a more realistic approach is to develop a simulation model. The simulation model helps us to explore some of the 'what if?' questions that require an answer for our plans for service development. Such simulations need delineation of our business rules, and so we need to decide, for example, what our normal procedures would be for dealing with a bank of photocopiers.

You might need to decide:

- Are all these photocopiers available during office hours – or do you keep one photocopier in reserve in case one breaks down?

- At busy times, do you restrict the amount of time an individual is allowed to spend photocopying?

Different situations will have different rules, but to obtain useful results from the simulation you need to think clearly about the business rules you operate – or could operate.

For more examples of the use of queuing theory to help manage demand in a way that suits you and your customers, you may consult some of the work done by the UK Government through the Modernisation Agency (*http://www .natpact.nhs.uk/demand_management/wizards/big_wizard/ menu.php*).

Their package stresses the importance of understanding the type of improvements that you wish to promote. For example, one method of dealing with a queuing problem might be to re-train staff, or allocate a different skill-mix of staff to deal with the queue, rather than simply open up another service point. It may be more productive for the library and information service manager to negotiate with other staff in the institution to ensure that library customers do not flood into the library at one time, all wanting to be served at once. If you are operating a virtual service, providing 'frequently asked questions' may relieve your helpdesk by channelling away the simple and easily answered queries, leaving them to deal with the more difficult and less frequent queries. Whatever you decide, it is probably a good idea to look at the process not just from the perspective of your staff, their skills and their stress levels, but also from the perspective of the customer and their likely 'journeys' through your service.

For many helpdesk situations, the service point may be acting as a router, directing enquirers to the most appropriate person to help them. This should mean that those who need priority attention get it, but if the enquiry is less urgent, users may be asked to wait. The difficulty in helpdesk situations that prioritise is that the users may have a different viewpoint on the priorities, and that time may be wasted in prioritising requests when it might be simpler and faster to treat the enquiries on a first come, first served basis. The Modernisation Agency work for the health sector notes that seeing the patient with the shortest procedure (p.74) first often means that the vast majority of patients are dealt with faster than under a first come, first served basis.

# Mapping what customers do in the library

For digital library services, transaction logs should provide some indication of what the users are doing and the routes they take through the digital library services. Managers may have more mundane reasons for asking what users are actually doing once they reach the library. Wear and tear on floor surfaces apart, there may be reasons for placing information about some new services at a point most users are likely to pass, and preferably pass slowly enough to

**Table 6.1**  Transition frequency table

|  |  |  |  |  |  | Row sums |
|---|---|---|---|---|---|---|
| Start | **(0)** | 17 | 30 | 5 | 0 | 52 |
|  |  | **S1** | 5 | 5 | 26 | 36 |
|  |  | 19 | **S2** | 3 | 13 | 35 |
|  |  | 0 | 0 | **S3** | 13 | 13 |
|  |  |  |  |  | **(Finish)** |  |
| Column sums |  | 36 | 35 | 13 | 52 |  |

*Source*: Ray et al. (2000)

realise that information is available. To analyse such data in a meaningful way, users need to be observed (difficult and intrusive) or interviewed on exit from the library. Ray et al. (2000) describe a method of tabulating the data as transition frequencies, for a simple example, with three different services (Table 6.1).

The table sums the total of these journeys. In total 17 people immediately went to S1, 30 people went to S2, and five people went to S3. The number of people exiting after S1 totalled 26 (12 of the people who visited S1 first, plus 14 of the 30 who first visited S2 and then S1). Nineteen people visited S1 after visiting S3. Three people in total visited S3 immediately after visiting S2, five people in total visited S3 after S1.

Such tables can provide an indication of the number of services accessed, and the general pattern of access. In the example provided in Ray et al. (2000), the least popular service accessed immediately after entering the library was advice and user education!

# Note

1. See, for example, Wisniewski (1997), Chapter 10.

# 7

# Conclusion

My aim in the book was not to provide yet another 'statistics for the terrified librarian', or 'quantitative methods in library management'. Rather, I hope that the chapters have assembled some techniques and ways of thinking about management problems that you may find useful. There is not really the space in the book to go into detail about some of the methods, but I have tried, where possible, to provide you with some references for further reading. Where there are few examples of direct application to library and information services, this usually reflects the fact that I did not find many useful examples in the published literature.

The mix of topics is an eclectic choice, and one that reflects what I have found useful or which I have used myself to solve some management problems. The order may not seem logical or very coherent, but I hope that within the topics you may find some nugget of information that sparks your interest, something that is novel to you. Most importantly, for those of you who dislike dealing with numbers, I hope that some of the suggestions give you ideas. I am sometimes despondent when I read some of the professional literature that does not seem to have progressed much in understanding of some basic quantitative library and information science problems over the past ten years. There are theoretical frameworks that can be borrowed from other disciplines – if management science has been doing this for years then there is every reason for library managers to take ideas from management sciences and apply them to the particular problems of managing the library or archive service of the future. Some of the problems faced by library managers are those shared by many managers. For example, the vast literature on performance measurement, audit and benchmarking offers many ideas that can be adopted relatively easily. Although other managers are concerned with the Pareto law problem, it seems to be particularly relevant for collection development and management in our field. Impact evaluation is a particular difficulty for the sector as it is so hard to get beyond the immediate impact assessment, but even an audit of that, suitably designed, can provide indications of longer-term impacts. The chapter on risk assessment showed how some techniques may help you decide what the best course of action is for you and your organisation, depending on the preferred attitude towards risk. It is so easy to suffer from planning blight, the impression that nothing can be done as a decision is awaited. Working through some of the likely outcomes for various scenarios can help decide whether or not the suggested route is really risky. Similarly, the final chapter suggested some ways of simulating and modelling the future. This is not the last word on such methods, but it was intended to indicate what is possible and where to make a start.

What is missing? One area that is not covered is systems thinking and systems analysis. Often this is where you need to start with the business problem, to ensure that you have identified the proper problem and that you do not rush to apply a technique to a problem that you think is there – but isn't. There is certainly scope for a good handbook on the application of business process modelling and systems analysis techniques to problems that library and information service managers face. However, this was beyond the scope of this particular book.

# Bibliography

Allen, T. J. (1977) *Managing the Flow of Technology: Technology Transfer and the Dissemination of Technological Information within the R&D Organisation*. Boston, MA: Massachusetts Institute of Technology.

Axelrod, R. (1997) *The Complexity of Cooperation: Agent–Based Models of Competition and Collaboration*. Princeton, NJ: Princeton University Press.

Barabási, A.-L. (2002) *Linked: The New Science of Networks*. Cambridge, MA: Perseus Publishing.

Bar-Ilan, J. (2005) 'What do we know about links and linking? A framework for studying links in academic environments'. *Information Processing & Management* 41: 973–86.

Belbin, R. M. (1993) 'Team roles at work'. Oxford: Butterworth-Heinemann.

Bensman, S. J. (2005) 'Urquhart and probability: the transition for librarianship to library and information science'. *Journal of the American Society for Information Science and Technology* 56(2): 189–214.

Bracke, P. J. (2004) 'Web usage mining at an academic health sciences library: an exploratory study'. *Journal of the Medical Library Association* 92 (4): 412–20.

Bradford, S. C. (1934) 'Sources of information on specific subjects'. *Engineering* 26: 85–6.

Bradford, S. C. (1948) *Documentation*. London: Crosley Lockwood.

Burrell, Q. L. (2005) 'Symmetry and other transformation features of Lorenz/Leikihler representations of informetric data'. *Information Processing & Management* 41: 1317–29.

Butler, M. A., Kingma, B. R. (editors) (1998) *The Economics of Information in the Networked Environment*. New York: Haworth Press.

Chapman, A. and Massey, O. (2002) 'A catalogue quality audit tool'. *Library Management* 23 (6/7): 314–24.

Cook, C. and Thompson, B. (2000a) 'Reliability and validity of SERVQUAL scores used to evaluate perceptions of library service quality'. *Journal of Academic Librarianship* 26(4): 248–58.

Cook, C. and Thompson, B. (2000b) 'Higher-order factor analytic perspectives on users; perceptions of library service quality'. *Library and Information Science Research* 22(4): 393–404.

Cook, C. and Thompson, B. (2001) 'Psychometric properties of scores from the web-based LibQUAL+ study of perceptions of library service quality'. *Library Trends* 49(4): 585–604.

Cooper, J., Spink, S., Thomas, R. and Urquhart, C. (2005) *Evaluation of the Specialist Libraries/Communities of Practice. Report for the National Library of Health*. Aberystwyth: Department of Information Studies; available at: *http://users.aber.ac.uk/cju* (last accessed: 17 January 2006)

Davies, R., Urquhart, C. J., Smith, J., Massiter, C. and Hepworth, J. B. (1997) 'Establishing the value of information to nursing continuing education: report of the EVINCE project'. BL RIC Report, 44. Boston Spa, Wetherby: British Library Document Supply Centre.

Egghe, L. (1985) 'Consequences of Lotka's law for the law of Bradford'. *Journal of Documentation* 41: 173–98.

Egghe, L. (2005a) *Power Laws in the Information Production Process: Lotkaian Informetrics*. Oxford: Elsevier.

Egghe, L. (2005b) 'Continuous, weighted Lorenz theory and applications to the study of fractional relative impact factors'. *Information Processing & Management* 41: 1330–59.

Egghe, L. and Rousseau, R. (2006) 'Classical retrieval and overlap measures satisfy the requirements for rankings based on a Lorenz curve'. *Information Processing & Management* 42: 106–20.

Fairthorne, R. A. (1969) 'Empirical hyperbolic distributions (Bradford-Zipf-Mandelbrot) for bibliometric description and prediction'. *Journal of Documentation* 25(4): 319–43; republished in *Journal of Documentation* 2005; 61(2): 171–93.

Haythornthwaite, C. (1996). 'Social network analysis: an approach and technique for the study of information exchange'. *Library and Information Science Research* 18: 323–42.

Hirshleifer, J. and Riley, J. G. (1992) *The Analytics of Uncertainty and Information*. Cambridge: Cambridge University Press.

Holström, J. (2004) 'The cost per article reading of open access articles'. *D-Lib Magazine* 10(1): DOI 10.1045/january2004-holmstrom.

Huang, C.-Y., Sun, C.-T. and Lin, H. C. (2005) 'Influence of local information on social simulations in small-world network models'. *Journal of Artificial Societies and Social Simulation* 8(4); available at: *http://jasss.soc.surrey.ac.uk/8/4/8.html* (last accessed 18 January 2006).

Kang, J. H. and Rouse, W. B. (1980). 'Approaches to forecasting demands for library network services'. *Journal of the American Society for Information Science* 31(4): 256–63.

Kingma, B. R. (1996) *The Economics of Information.* Englewood, Co: Libraries Unlimited, Inc.

Kinnucan, M. (1993) 'Demand for document delivery and interlibrary loan in academic settings'. *Library and Information Science Research* 15: 355–74.

Loria, A. and Arroyo, P. (2005) 'Language and country preponderance trends in MEDLINE and its causes'. *Journal of the Medical Library Association* 93 (3): 381–5.

Martin, S. (2003) 'Using SERVQUAL in health libraries across Somerset, Devon and Cornwall'. *Health Information and Libraries Journal* 20(1): 15–21.

Marshall, J. G. (1992) 'The impact of the hospital library on clinical decision making: the Rochester study'. *Bulletin of the Medical Library Association* 78(2): 180–7.

Maynard Smith, J. (1982) *Evolution and the Theory of Games.* Cambridge: Cambridge University Press.

Moreno, A. (2001) 'Enhancing knowledge exchange through communities of practice'. *Aslib Proceedings* 53(8): 296–308.

Nicholas, D., Huntington, P. and Williams, P. (2001) 'Establishing metrics for the evaluation of touch screen kiosks'. *Journal of Information Science* 27(2): 61–71.

Nicholas, D., Huntington, P. and Williams, P. (2004) 'Re-appraising information behaviour in a digital environment: bouncers, checkers, returnees and the like'. *Journal of Information Science* 60(1): 24–43.

Nicholas, D., Huntington, P., Dobrowolski, T., Rowlands, I., Jamali M. R., Polydoratou, P. (2005) 'Revisiting "obsolescence" and journal article "decay" through usage data: an analysis of digital journal use by year of publication'. *Information Processing & Management* 41: 1441–61.

Nicholson, S. (2003) 'Bibliomining for automated collection development in a digital library setting: using data mining to discover Web-based scholarly research works'. *Journal of the American Society for Information Science and Technology* 54(12): 1081–90.

Paulos, J. A. (1988) 'Statistics, trade-offs and society' *Innumeracy: Mathematical Illiteracy and its Consequences*. London: Penguin.

Payne, N. and Thelwall, M. (2005) 'Mathematical models for academic webs: linear relationship or non-linear power law'. *Information Processing & Management* 41: 1495–510.

Ravid, G. and Rafaeli, S. (2004) 'Asynchronous discussion groups a small world and scale-free networks'. *First Monday* 9(9); available at: *http://firstmonday.org/issues/ issue9_9ravid/index.html* (last accessed: 17 January 2006).

Ranganathan, S. R. (1963) *The Five Laws of Library Science*. Bombay: Asia Publishing House.

Ray, K., Heine, M., Winkworth, I. (2000) 'What users actually do: a study of service-seeking behaviour in an academic library'. *Library & Information Research News* 24(76): 18–25.

Rousseau, R. (2005) 'Robert Fairthorne and the empirical power laws'. *Journal of Documentation*, 61(2): 194–202.

Shan, S. (2005) 'On the generalized Zipf distribution. Part I'. *Information Processing & Management* 41: 1369–86.

Tversky, A. and Kahnemann, D. (1974) Judgment under uncertainty: heuristics and biases. *Science* 185 (27 September), 1124–31.

Tversky, A. and Kahnemann, D. (1981) The framing of decisions and the psychology of choice *Science* 211 (30), 453–58.

Urquhart, C. (2002). 'Applications of outsourcing theory to collaborative purchasing and licensing'. *VINE* 32 (4): 63–70.

Urquhart, C., Spink, S. and Thomas, R. (for National Library for Health) (2005) *Assessing Training and Professional Development Needs of Library Staff.* Aberystwyth: Department of Information Studies, UWA; available at: *http://users.aber.ac.uk/cju* (last accessed: 18 January 2006).

Urquhart, C. J. and Hepworth, J. B. (1995) *The Value to Clinical Decision Making of Information Supplied by NHS Library and Information Services.* British Library R&D Report 6205. Boston Spa, Wetherby: British Library Document Supply Centre.

Watts D. J. and Strogatz, S. H. (1998) 'Collective dynamics of 'small-world' networks'. *Nature* 393, 440–2.

Webb, J. and Galloway, L. (2000) 'The voice of the process: process management and performance measurement of collection development activities in a new university library'. In *Proceedings of the 3rd Northumbria International Conference on Performance Measurement in Libraries and Information Services, Value and Impact,* 27–31 August 1999. Newcastle upon Tyne: Information North for School of Information Studies, University of Northumbria, pp. 125–130.

Wisniewski, M. (1997) 'Decision making under uncertainty' *Quantitative Methods for Decision Makers.* London: Pitman Publishing.

Wolfram, D. (2003) *Applied Informetrics for Information Retrieval Research.* New Directions in Information Management no. 36. Westport, CT: Libraries Unlimited.

Yeoman, A., Urquhart, C., Cooper, J., and Tyler, A. (2001) *The Value and Impact of Virtual Outreach Services: Report of the VIVOS Project.* Library and Information Commission report 111. London, Resource: The Council for Museums, Archives and Libraries.

# Index